T0215294

Biotech Juggernaut

Biotech Juggernaut: Hope, Hype, and Hidden Agendas of Entrepreneurial BioScience relates the intensifying effort of bioentrepreneurs to apply genetic engineering technologies to the human species and to extend the commercial reach of synthetic biology or "extreme genetic engineering." In 1980, legal developments concerning patenting laws transformed scientific researchers into bioentrepreneurs. Often motivated to create profit-driven biotech start-up companies or to serve on their advisory boards, university researchers now commonly operate under serious conflicts of interest. These conflicts stand in the way of giving full consideration to the social and ethical consequences of the technologies they seek to develop. Too often, bioentrepreneurs have worked to obscure how these technologies could alter human evolution and to hide the social costs of keeping on this path. Tracing the rise and cultural politics of biotechnology from a critical perspective, *Biotech Juggernaut* aims to correct the informational imbalance between producers of biotechnologies on the one hand, and the intended consumers of these technologies and general society, on the other. It explains how the converging vectors of economic, political, social, and cultural elements driving biotechnology's swift advance constitutes a juggernaut. It concludes with a reflection on whether it is possible for an informed public to halt what appears to be a runaway force.

Tina Stevens, Ph.D., is Lecturer Emerita at San Francisco State University, Department of History. She is a co-founder of Alliance for Humane Biotechnology, and the author of *Bioethics in America: Origins and Cultural Politics* (Johns Hopkins University Press, 2000).

Stuart Newman, Ph.D., is Professor of Cell Biology and Anatomy at New York Medical College where he studies developmental and evolutionary biology. He was a founding member of the Council for Responsible Genetics and is co-author of *Biological Physics of the Developing Embryo* (Cambridge, 2005). He is editor of the journal *Biological Theory* (Springer).

Stevens' and Newman's keen insights are grounded both in a long view – putting biotech and its societal implications in broad historical context – and a close-up one, as they recount their personal encounters with the biotech juggernaut in scientific, legal, policy, and advocacy settings. Their stories demonstrate the sweeping commercialization of the biotechnology enterprise, its routine conflicts of interest, and its tendency to exaggerate benefits and minimize hazards. At a time when headlines blare claims of gene-edited babies, their challenge and their guidance are indispensable.

Marcy Darnovsky, Executive Director of
the Center for Genetics and Society

Tina Stevens and Stuart Newman pull the curtain back on the biotech industrial complex, revealing how its toxic combination of clever marketing and political and financial muscle is perpetuating a branch of science at odds with the public interest. *Biotech Juggernaut* is also the story of how a small group of dedicated activists is fighting back. By the end of this book, you'll want to join them.

Jeremy Gruber, Former President of
the Council for Responsible Genetics

Stevens and Newman's deep familiarity with the events and issues is on clear display in *Biotech Juggernaut*. Many books address issues in bioscience, and yet fall short of the authors' rich historical and institutional analysis and their especially illuminating case studies. Even fewer are written with such enlightening critical perspective.

Elaine Draper, Professor of Sociology,
California State University, Los Angeles

Biotech Juggernaut is a compelling exposé of high stakes discoveries in genetics with unrealistic expectations pushed prematurely into clinical applications. The authors take on the quintessential hubris of scientists who thumb their nose at evolution, taking it upon themselves, in the face of broad public opposition, to redesign the human genome opening the door to eugenics.

Sheldon Krimsky, Lenore Stern Professor of Humanities &
Social Sciences and Adjunct Professor of
Public Health & Community Medicine,
Tuft University, and author of *GMOs Decoded*

Biotech Juggernaut

Hope, Hype, and Hidden Agendas of Entrepreneurial BioScience

Tina Stevens and Stuart Newman

Routledge
Taylor & Francis Group

NEW YORK AND LONDON

First published 2019
by Routledge
52 Vanderbilt Avenue, New York, NY 10017

and by Routledge
2 Park Square, Milton Park, Abingdon, Oxon, OX14 4RN

Routledge is an imprint of the Taylor & Francis Group, an informa business
© 2019 Taylor & Francis

Cover Image by Roberto Matta, "L'ame du Tarot de Theleme 1" © 2018 Artists Rights Society (ARS), New York / ADAGP, Paris. Courtesy of Denis Bloch Fine Art, denisbloch.com

Library of Congress Cataloging-in-Publication Data
A catalog record for this title has been requested

ISBN: 978-1-138-04319-0 (hbk)
ISBN: 978-1-138-04323-7 (pbk)
ISBN: 978-1-315-17326-9 (ebk)

Typeset in Minion
by Swales & Willis Ltd, Exeter, Devon, UK

For Peggy Stevens
in memoriam
Who, long ago, held a shell to my ear promising I would hear the sea.
and
for Stephen Shmanske
Who shows me, daily, the wisdom of such fascination and delight.
– Tina Stevens

For Jura
– Stuart Newman

Contents

Acknowledgments

When our paths crossed in 2004, we did not then know how parallel awakenings experienced decades before had marked out that juncture. Only in 2016 did we discover that one of Stuart's heroes had been one of Tina's mentors. Carolyn Merchant's masterful treatise, *The Death of Nature*, galvanized our awareness, leading ultimately to the critique constituting this volume. For that formative quickening and for Carolyn's continuing counsel we are deeply grateful.

Sadly, it is *in memoriam* only that we may acknowledge heartfelt gratitude to two colleagues *cum* mentors. MIT historian of science Charles Weiner's impeccable chronicling and analysis of the Responsible Science Movement informed our effort throughout. McGill University epidemiologist, Abby Lippman was a founding member of the Council for Responsible Genetics, a founding director of Alliance for Humane Biotechnology, and an early voice calling out the perils of human genetic engineering. Their commitment to bridging the gulf between academics and activism continues to inspire.

We are especially grateful to Diane Beeson, Marcy Darnovsky, Elaine Draper, Richard Hayes, Becky McClain, Marsha Saxton, Pete Shanks, and Jim Thomas. They generously brought their expertise to bear on chapter drafts. Some also brought the inspiration of enlightened, dedicated activism.

At various points we benefitted from support or advice from Chicago-Kent College of Law's Lori Andrews, UC Berkeley's professor emeritus Troy Duster, Council for Responsible Genetics' Former President Jeremy Gruber, David King of Human Genetics Alert, U.K., and Tufts University's Sheldon Krimsky. We remain grateful, recognizing how their counsel, as well as insights provoked by their scholarship and publications, energized our undertaking. Judy Norsigian's life's work of scientifically informed advocacy has been a paragon for us as it has been for many others.

Tina offers singular thanks to Stuart Newman. His courage and tenacity in insisting that rigorous scientific study can and must be rigorously socially responsible undergirded this project. His assurance that scientific complexities can and should be rendered broadly understandable made it possible.

His generous collaboration in parsing such complexities pushed a cheeky proposal into a completed reality.

Her abiding affection for Rosann Greenspan, Patti Martin, and James E. Stevens includes special gratitude for their unwavering encouragement and highly critical reads that often improved the text. For valued feedback, support, or simply forbearing a friend and family member's preoccupation for too many years, tender and sincere thanks to Pat Conolly, Sherry Katz, Steve Leikin, Kyle Livie, Joan C. Ryan, Teri Reynolds Stevens, Gloria Jeanne Stevens, Richard O. Stevens, William A. Stevens, Lisa Zemelman, and to the GenXers and Millennials who fill out the family tree. Enduring gratitude for collaborations with Diane Beeson. Diane's integrity often led the way to effective action that her scholarship infused with authority.

Her deepest thanks are for Stephen Shmanske. He insisted on this project. He encouraged every passage with the intent listening of his superbly critical ear. With boundless love and support, he coaxed its completion.

Stuart extends his gratitude to Tina Stevens, who conceived this project and recruited him to it, and ceaselessly turned back the passages he wrote until they communicated their ideas nearly as lucidly as hers. Tina's role, and Diane Beeson's and Abby Lippman's, in several of the episodes recounted here, have been invigorating examples of, and models for social engagement by academics in the ethics of biotechnology and medicine.

Stuart also acknowledges companions, friends, and colleagues who have informed and enriched his understanding of the scientific, political, and moral dimensions of the subjects discussed in this book. These include most importantly Ramray Bhat, Malcolm Byrnes, Felipe Cabello, William Hurlbut, Jonathan King, Marta Linde-Medina, Gerd Müller, Vidya Nanjundiah, Jura Newman, Jeremy Rifkin, and James Siegel.

We thank Dean Birkenkamp and Tyler Bay, our editors at Routledge, for their encouragement and attention to this project, and Jane Fieldsend for her skilled copyediting. We are also grateful to Todd Leibowitz of the Artists Rights Society, New York, and Dionne Wilson and Denis Bloch, of Denis Bloch Fine Art, Beverly Hills, California for, respectively, helping us obtain permission to use the book's cover image, and providing the picture.

Prologue
A Shared Encounter

juggernaut (noun):

an overwhelming or unstoppable force
www.thefreedictionary.com/Juggernaut

anything that exacts blind devotion or terrible sacrifice
www.collinsdictionary.com/us/dictionary/english/juggernaut

The summer of 2004, Tina received an ominous phone call from the office of California's Secretary of State. Pleasant sounding enough, the staffer wanted to confirm her address and the correct spelling of her name, volunteering that such inquiries usually meant that formal legal action was about to be filed. Legal action? What kind of legal action? About what exactly? The fact that this was the Secretary of State's office calling sunk in and Tina understood that this must have something to do with the rebuttal that she and two others had filed urging voters to register a NO vote against state Proposition 71, the California Stem Cell Research and Cures Initiative. Judy Norsigian, Executive Director of Our Bodies Ourselves, Francine Coeytaux of the Pacific Institute for Women's Health and Tina, a historian of bioethics, co-filed the rebuttal as part of a loosely connected group of concerned parties that eventually would call itself the ProChoice Alliance Against Prop 71 (PCA). For now, they were just a handful of women's health advocates, academics, and public interest groups, networking in earnest since learning that a significant threat to women's health lurked at the heart of Prop 71. The proposition was strongly supported by most liberals and progressives, in part because opposition to embryonic stem cell research was widely seen as a position associated with President George W. Bush and the religious right. Most liberals and progressives seemed unaware, and the proposition campaign did not reveal, how the initiative would facilitate experimental human cloning research, a type of cloning that required women's eggs, large untold numbers of them. But could this call *really* be about that? Who could mobilize so swiftly – the rebuttal had been filed just days before – and on what grounds? Is someone trying to suppress our dissent? And if so, who? The staffer couldn't or wouldn't explain more. The answer, confirmed a few days later, pulled back the curtain on the iron jaw of a juggernaut: two scientists and the financier-author of the proposition were petitioning to have particular statements of fact in the threesome's rebuttal blocked from appearing in the Voter's Guide. These petitioners possessed intimidating clout including access to an astonishing multimillion dollar war chest, more than sufficient for going to court to halt disclosures they asserted were "false and misleading." How would it be possible to fight their allegations? Tina reached out to Stuart who, as a developmental biologist who had testified before the U.S. Congress about human cloning issues, might be willing to assist in contesting the charges. Stuart was also a founding member of the Council for Responsible Genetics, which presented the first statement denouncing human germline manipulation of the modern molecular genetics era (Appendix A). Helping to defeat these charges was our first shared encounter with some of the social-political forces of biotechnology. The Voter's Guide went to press with our contributions essentially unchanged.

Our book, *Biotech Juggernaut: Hope, Hype, and Hidden Agendas of Entrepreneurial BioScience* tells this story (in Chapters 3 and 4) as part of its account of some of the more troubling aspects of the rise of bio-entrepreneurialism, especially as they relate to human genetic engineering. We relate some of the arguments and strategies used to promote, and sometimes to conceal, the transformation of all living things, including ourselves, to new purposes. Biotechnologies we discuss, as they relate to manipulating the human species, include cloning, "three-parent" embryos, gene editing, synthetic genome creation, and human–animal embryonic combination. We shed light on how biotechnology is straying well beyond the border separating homo sapiens from "GMO Sapiens" (Knoepfler, 2015). In describing perceived and real conflicts of interest in the biotech industry, we hope to spark needed conversation about who should be the gatekeepers and framers of public discussion about the wisdom of crossing over that crucial boundary. Currently, bioentrepreneurs themselves too often monopolize this important role.

In May 2016, for example, 150 scientists, entrepreneurs, and lawyers met at Harvard University to discuss creating a completely synthetic human genome for insertion into a cell line. This invitation-only event was closed to the media. Executive Director for the watchdog organization Center for Genetics and Society (CGS), Marcy Darnovsky, characterized the "semi-secret meeting . . . making plans about synthesizing the human genome" to be "a new low in scientific accountability" (CGS Comment, May 13, 2016).

In the U.S. and around the globe, public opinion demonstrates pervasive revulsion at the prospect of genetically modifying the human species (CGS, 2014, 2015; Kolata 2016). Yet, industry-led discussion leaves the door wide open to normalizing just that. In December 2015, the national scientific academies of the U.S., U.K., and China convened a "gene editing summit" in Washington DC. Although the event received wide media attention, its participant roster included few critics. In their official statement, however, summit officials navigated a narrow line, disingenuously employing some of the critics' language, but in a way that could be, and was, later disavowed. They stated, "[i]t would be irresponsible to proceed with any *clinical* use of [human] germline editing, unless and until . . . there is broad societal consensus about the appropriateness of the proposed application" (ibid). In other words, despite strong social disapproval and laws in dozens of countries around the world, prohibition of human germline (egg and sperm) gene editing was not seriously considered a viable option. By February 2017 the U.S. National Academies of Science, and of Medicine were ready to dispense with any such coy formulations, asserting that "[c]linical trials using heritable germline genome editing should be permitted" (National Academies, 2017, p. 102). Finally, while this book was in proofs, a scientist from China announced at a second gene editing summit in Hong Kong in November 2018 that he had performed such editing on twin infant girls (Begley, 2018). He claimed that the National Academies report had justified his doing so.

Why? How did we come to the point of sliding past the line most people do not want transgressed? National debate about key biotechnological developments deserves widening. What's at stake for our species deserves fresh, urgent clarification. The analysis we offer here acknowledges the legitimate promise of biotechnology and the well-intentioned motivations sometimes driving its development. But we also call attention to powerful influences within the professional cultures of science and bioentrepreneurialism, including profit-seeking and career-building, that foster exaggerating benefits, masking hazards, and obscuring the extent to which human genetic modification is already very much underway. We conclude with a query: is social activism about human genetic technologies possible? In the hope that it may be, we offer suggestions for how concerned citizens can become informed participants in securing a human future.

References

Begley, Sharon, "Claim of CRISPR'd Baby Girls Stuns Genome Editing Summit," *STAT*, November 26, 2018: www.statnews.com/2018/11/26/claim-of-crispred-baby-girls-stuns-genome-editing-summit/

Center for Genetics and Society, "Summary of Public Opinion Polls," 2014: www.geneticsandsociety.org/article.php?id=401

Center for Genetics and Society, "Human Germline Modification Summary of National and International Policies," June 2015: www.geneticsandsociety.org/downloads/CGS_Global_Policies_Summary_2015.pdf

Knoepfler, Paul, *GMO Sapiens: The Life-Changing Science of Designer Babies*, Hackensack, NJ: World Scientific, 2015.

Kolata, Gina, "Building a Better Human with Science? The Public Says, No Thanks," *New York Times*, July 26, 2016: www.nytimes.com/2016/07/27/upshot/building-a-better-human-with-science-the-public-says-no-thanks.html?_r=0

National Academies of Sciences, Engineering, and Medicine, *Human Genome Editing: Science, Ethics, and Governance*, Washington, DC: The National Academies Press, 2017, 328pp.

1
Introduction
The Biotech Juggernaut

Could not the race of men be similarly improved? Could not the
undesirables be got rid of and the desirables multiplied?
> (Sir Francis Galton, Eugenics theorist (contemplating applying
> cattle breeding methods to humans), 1908
> (Galton, quoted in Grogan, 2014, p. 94))

Ultimately [the new biology] could diagnose, then specify, the actual
DNA composition of ideal man.
> (Joshua Lederberg, Nobel Laureate,
> Stanford molecular biologist from,
> "Molecular Biology, Eugenics and Euphenics" (Lederberg, 1963))

What is it that makes germline manipulation of humans special?
It's . . . our perception of ourselves. If we feel that we can change any
aspect of ourselves, where do you begin and where do you stop? and
who sets those rules?
> (George Church, Harvard University Professor, "The augmented
> human being," interview in *The Edge* (Church, 2016))

What is biotechnology? The industry's largest trade association,
Biotechnology Innovation Organization (BIO), describes biotechnology
as harnessing "cellular and biomolecular processes to develop technologies
and products." BIO claims a 6,000-year history for biotechnology, in which
the biological processes of microorganisms have been used to make "useful
food products, such as bread and cheese, and to preserve dairy products"
(www.bio.org/). Seen through this unfocused lens, there is no significant
historical difference between ancient methods of cheese production and
the genetic modification of human embryos in the lab.

Exposing the misleading nature of this stylized history is the fact that the
term "biotechnology" wasn't even coined until 1917 (Bud, 1994), and it was
hardly in use until a dramatic increase in the 1970s. Ten sporadic references

in the scientific literature in the entire quarter century from 1947 to 1971 grew to over 200,000 in the past 40 years. Countless mentions in popular media followed suit. This steep climb paralleled the explosion of other indicators: educational programs, conferences, journals, trade associations, unique financing arrangements, new corporations, and ethics panels all dedicated to "biotech."

The Rise of the Biotech Industry

In fact, something bold and unprecedented had happened within bioscience: in the early 1970s researchers learned to recombine **DNA**. Removing a **gene** from one organism, they could "splice" it into another. Short-order laboratory transactions could now induce biological transformations previously occurring only over millennia or not at all. Historian of Science Susan Wright makes plain that for contemporaries, "controlled linking of genes from different species was seen . . . as a novel and remarkable achievement." Wright quotes one British scientist in 1974: "For the first time, there is now available a method which allows us to cross very large evolutionary barriers and to move genes between organisms which have never had genetic contact" (Wright, 1994). For many at the time "genetic engineering" was an audacious challenge to the moral order and to the safety of nature's vast, exceedingly unhurried experiment of evolutionary creation.

This rapid-paced, human directed laboratory remixing triggered the emergence of biotechnology. At once a scientific field of study and a technological platform, biotechnology seamlessly merges basic and applied science. In this blended arena, the desire for commercialized bio-products dictates research agendas as unbounded as the human imagination, from creating microorganisms capable of eating slicked oil to designing human embryos to glow green. The rise of biotechnology altered scientific practice and professional culture and brought in its wake social, legal, political, and moral transformations.

Some 40 plus years since biotech's emergence, critical reflection on the roots, ramifications, prospects, and promises of this highly consequential field is overdue. The assessment we offer here is not about the underlying science driving biotech per se, although scientific explanations are woven throughout. Instead, this is an account of what we call the biotech juggernaut: the converging vectors of economic, political, social, cultural, and ethical elements driving biotechnology's swift advance, especially in regard to applications to human biology. Private venture capital, originally reticent to embrace biotech, viewing its sanguine promises insufficiently grounded, eventually came to embrace it. But in certain cases a poorly informed public, riding high on hope and promises, also opened its coffers to fuel the industry. Biotechnology's history is marred by ethical abuses, clinical failures, hidden agendas, false promises, hype, and private bonanzas snatched at public expense. Yet, despite a record warranting at least as much caution as enthusiasm, by 2017 venture capital investment in the biotech sector reached more than $10 billion (PitchBook, 2017) on top of more than $130 billion in biotech facility investments by the pharmaceutical industry around the world (Kreuger, 2018). Biotech has indeed become a juggernaut.

Ethics, Dissent, and the Dream of Responsible Science

Genetic laboratory discoveries would not have become the foundation of a twenty-first century "bio-economy" without the continuous quelling of concern over their social and ethical implications. Unveiling the mysteries of organismal development and empowering humankind to engineer novel life forms provoked anxiety at least as far back as H.G. Wells's 1896 science fiction fantasy, *The Island of Dr. Moreau.* Aldous Huxley's biologically sophisticated and socially ominous dystopian 1931 novel *Brave New World,* forecasted chilling social outcomes. By the 1950s, World War II's **eugenics** inspired extermination death camps gathered a dark cloud of reality over excitement that accompanied the discovery of DNA's structure and the implied possibilities of its function. Historian Charles Weiner described a relatively reflective "pre-recombinant DNA period" from the 1960s to the mid 1970s when some scientists, anticipating ethical interest in the genetic interventions they knew were coming, offered fretful reflection in both professional and, less frequently, public venues (Weiner, 1999).

A number of these scientists had been affected deeply by the perceived moral shortcomings of the atomic scientists who had worked on the so-called Manhattan Project that produced the atomic bomb. Life science researchers wanted to escape the kind of condemnation leveled at the physical scientists who had failed to consider the long-term moral implications of their research or to involve the public. Those scientists were believed to be complicit in the loss of control over that powerful technology, not having secured the ability to shield it from ill use by vested and political interests.

Biochemist and Nobel Laureate Arne Tiselius linked the atomic explosions at the conclusion of World War II to the forecasted power of genetic discoveries. When he rose to address the audience at the 60th anniversary of the Nobel Prize in 1961 he brought attention to the moral obligations of scientists. They must, he urged, attend to how their work facilitated applications in weapons of mass destruction and, adding quickly, "the search into the very basis of the life process." "[I]t is possible," he continued, "many even think probable," that biological research "will lead to methods of tampering with life, of creating new diseases, of controlling the psyche, of influencing heredity, even perhaps in certain desired directions." He warned gravely: such tampering could "result in a still more refined and perhaps still more dangerous way of abusing the results of research than that implied in the instruments of mass destruction." He saw the need "for an international moral code governing the use of scientific results" (cited in Weiner, 1999, p. 52).

The Specter of Eugenics

Of the concerns Tiselius marked, one in particular resonated chillingly: "influencing heredity." Beginning in the 1890s a broad spectrum of scientists, academics, and legal and civic leaders had attempted to influence heredity by embracing "eugenics." The goal of eugenics was to "improve" the human race. Eugenicists advocated the adoption for human use of techniques developed in animal breeding to increase heritable characteristics considered desirable. Few Americans today seem to be familiar with the eugenics movement in the U.S. and Britain. But historically, the agenda was not a socially marginalized one (Comfort, 2014). Many scientists at the nation's premier universities, Supreme Court justices, university presidents, authors (e.g., H.G. Wells and George Bernard Shaw), a host of prominent citizens (e.g., Alexander Graham Bell) and foundations and institutes (e.g., Rockefeller Foundation and Carnegie Institution), and national leaders including Theodore Roosevelt, Herbert Hoover, and Winston Churchill supported eugenic ideas and programs. International Eugenic Congresses were held in 1912, 1921, and 1932, the first in London, the second and third in New York, so that scientific and civic leaders could consider how best to "eliminate the unfit" and "improve" human heredity. Eugenic promotion permeated all levels of society, finding expression in professional journals and college courses as well as state fair attractions. "Fitter Family Contests" and eugenic exhibits were popular features at state fairs. Competitions took place in the "human stock" division. As one contest brochure related, "The time has come when the science of human husbandry must be developed, based on the principles now followed by scientific agriculture, if the better elements of our civilization are to dominate or even survive" (Kevles, 1985, pp. 61–62.) A prevalent U.K. race betterment poster from the 1930s depicted a robust male laborer and urged emphatically that, "Only *Healthy* Seed Must be Sewn! Check the Seeds of Hereditary Disease and Unfittness by Eugenics" (Hall, 1990). In 1927, the U.S. Supreme Court found, in *Buck vs. Bell*, that Carrie Buck, her mother, and her daughter were "feeble minded." Judging that society can prevent the "manifestly unfit from continuing their kind," the court upheld state eugenic sterilization laws. Writing for the 8 to 1 majority, Justice Oliver Wendell Holmes, Jr. famously declared that, "three generations of imbeciles are enough" (*Buck vs. Bell*). Thus, cloaked by sanctity of law, physicians across the nation forcibly sterilized tens of thousands of citizens deemed unfit to reproduce. Apart from those considered mentally inadequate (and appallingly, often conflated with them) were disproportionate numbers of Black women and Latinas. Sterilizations continued in government sanctioned venues, including county hospitals in California, up through the 1970s (Stern, 2016).

Only after World War II was it understood that the Nazi death camp atrocities, promulgated by a *fuhrer* enamored of eugenic imperatives contrived in the U.K. and U.S., were the most brutal expression of a genetic wish to eliminate "undesirables" from a gene pool. The anticipated genetic revolution could, if left unguided by moral reflection and unlimited by ethical boundaries, encourage a science-spurred version of similar eugenic outcome. Later dubbed "techno-eugenics" by the scholar and co-founder of the Center for Genetics and Society, Richard Hayes (Hayes, 2000), it rested upon the unfounded conceit of what came to be a central dogma of the new genetics: that there were straightforward and readily discernable relationships ("mappings") between genes and traits (See Chapter 2 for a discussion of this fallacy). This doctrine raised a question and presented a possibility: if one or a few genes could be causally associated with specific characteristics would it not be possible to one day engage the eugenic strategy of eliminating "undesirable" qualities from the gene pool and promoting those desired?

Scientists pondered the ethical limits of human genetic engineering in advance of the expected technical ability, emerging finally in the early 1970s, to perform genetic interventions. Chastened by the eugenic underpinnings of the Third Reich's "final solution," eugenic impulses within the postwar scientific community subsided. The scientific journal, *Annals of Eugenics*, founded in 1925, for example, changed its name in 1954 to *Annals of Genetics*, the title it still bears. (Astonishingly, the British Eugenics Society became the Galton Institute only in 1989.) But eugenic impulses did not vanish. Alternative names were found – "euphenics" or "germinal choice" and, thus disguised, eugenic ideas lurked furtively in scientific papers and conferences (Stevens, 2003). But the influence of the atomic experience remained visibly galvanizing. Dramatic imagery at one 1970 conference featured a mushroom cloud intertwined with the double helix (Weiner, 1999, p. 55).

Scientists who worked on the Manhattan Project and committed thereafter to warning the public of technological perils founded the Federation of American Scientists in 1945. During the 1960s and '70s, vigorous additional influences amplified the undimmed lessons of the atomic scientists. The environmental and anti-war movements spawned agitating sensibilities about the potential misuses of scientific knowledge, especially over possible military applications of chemical and biological warfare (Wright, 1994, p. 124). New groups questioning the ends of science and promoting the social responsibilities of scientists emerged. Ranging from moderate to radical, these groups included the Society for Social Responsibility in Science, the Scientists' Institute for Public Information, the Medical Committee for Human Rights, MIT's Science Action Coordinating Committee, and

Scientists and Engineers for Social and Political Action, known eventually as, Science for the People. Positions ranged from recognizing a need to alert the public about the hazards of potential laboratory accidents and adverse ecological effects to opposing how science often served to bolster existing social and economic relations. And, for a broad range of experts and concerned citizens the fundamental question remained not how the new technologies *could* be implemented but whether they *should* be (Krimsky, 1985, pp. 16–22).

The essential role that a presumptively neutral science had played in making possible the atomic devastation of human life and the dawn of the arms race remained for many scientists an undiminished cautionary tale. Robert Pollack, a cancer researcher at Cold Spring Harbor Laboratory made the resonating impact of atomic allusions explicit when in 1973 he asserted that, "We're in a pre-Hiroshima condition. It would be a real disaster if one of the agents now being handled in research should in fact be a real human cancer agent" (Wright, 1994, p. 127). Janet Mertz, who worked in Paul Berg's lab as a graduate student in 1970 later shared that she "started to think in terms of the atomic bomb and similar things. I didn't want to be the person who went ahead and created a monster that killed a million people" (Krimsky, 1985, p. 31). Sentiments expressed in 1976, by chairman of the Division of Biology at the California Institute of Technology, Robert Sinsheimer, echoed sensibilities of the postwar anticipatory era:

> the description of life in molecular terms provides the beginning of a technology to reshape the living world to human purpose, to reconstruct our fellow life forms – each as are we, the product of three billion years of evolution – into projections of human will. And many are profoundly troubled by the prospect.
>
> (cited in Wright, 1994, p. 113)

But the enduring moral lessons of the atomic legacy chaffed against what were, for some, opposing worries: a fear that public "overreaction" might result in loss of funding and control over research agenda. Reticent to discuss genetic engineering publicly, some conferees decided against official reporting of ethical deliberations. In the late 1960s, congressional hearings considered institutionalizing a national committee on health, science, and society with an emphasis on emerging medical technologies, including genetic engineering. Some biologists, increasingly reluctant to participate and more defensive about their research, fashioned a public stance that functions to this day: worries about ills to come from genetic interventions in human life were premature speculations. Such unfounded fears, they argued, served only to delay imminent cures for deadly diseases (Weiner, 1999, pp. 56–57). This artful position obscured how the earliest concerns were aired first by scientists themselves, including worries over misleading the public with hyped talk of the promise of cures. The manipulation of public expectation was of particular concern to some critics in the 1970s and '80s as rudimentary research into gene splicing, first in microbes and plants and then in animals, led to increasingly insistent proposals to remake human biology.

Technological Pragmatism Replaces Moral Concern

As researchers began closing in on the laboratory techniques that would make genetic engineering a reality, the wide moral scope of the "anticipatory era" in which there was open and accepted speculation about social consequences narrowed. What was of most concern to those in the field was lab safety. In 1970, Stanford scientist Paul Berg planned to introduce a tumor-causing gene into a bacterium. In 1970 scientists revealed newly discovered "restriction enzymes" that could cut strands of DNA at specific points to combine them with DNA of other organisms. These developments raised alarm over the biohazards of creating novel organisms. When, in 1974, some bioscientists called for a moratorium on genetic engineering research, it was not inspired by the expansive ethical framing characteristic of the anticipatory era. At the now famed 1975 Asilomar conference, sometimes viewed as a milestone of responsible science, the opening remarks of scientist David Baltimore exemplified a much constricted field of moral vision:

> Although I think it's obvious that this technology is possibly the most potent potential technology in biological warfare, this meeting is not designed to deal with that question . . . We are here . . . to balance the benefits and hazards right now and to design a strategy which will maximize the benefits and minimize the hazards for the future.
>
> (Weiner, 1999, p. 58)

Whether to proceed was not the chief concern, only how. Indeed, in November 1974, evidently not perceiving any need to wait for an end to the moratorium, two California biologists filed for a patent on the technique developed to recombine DNA. The 1974 moratorium and the 1975 Asilomar conference helped extricate the double helix from its cautionary association with the mushroom cloud.

A group of academic scientists, policy analysts, and labor activists, some with past affiliations to Science for the People consolidated around a 1977 battle between Harvard University and the government of Cambridge, Massachusetts on siting a gene splicing laboratory. Although safety was the friction point in Cambridge, eventually the discussions of the group developed into a broader critique of the role of new genetics in society, and in the early 1980s led to the founding of the Council for Responsible Genetics

(CRG), the first public interest group with a focus on the full range of these emerging technologies (Krimsky, 1985; Wright, 1994). Its founding echoed the responsible science critique of the earlier part of the century. But it was an exception, and its strong stance against genetic engineering of humans (Lippman et al., 1993; Billings et al., 1999), would eventually wane.[1] After Asilomar, as bio-researchers with a financial stake in developing genetic technologies set aside anxieties over how to avoid destructive uses of the powerful genetic technologies they were developing, the influence of the responsible science movement began to dim. (Eventually, even the CRG would relinquish its early call for a ban on genetic engineering of humans.) By narrowing the moral landscape that first incubated concern, genetic engineers ensured that any new regulatory policies would be developed out of the public eye.

* * * * * * * *

Claiming potential benefit and disparaging the concerns of non-experts became and remains a winning strategy for biotech development. Indeed, "promise" is a hallmark of biotech's continuing success as a social institution: the potential of biotechnology for finding cures, improving the quality of life, sometimes improving (according to the lights of the developers) life itself. The poster child for such promise is the production of human insulin in bacteria by gene insertion or splicing in 1978. Where diabetic patients had previously depended on animal insulins, which were more costly and potentially allergenic, they could now be treated with a product indistinguishable from that produced by members of their own species. Recently, advances in the field of gene therapy also constitute significant fulfillment of biotech's promise. These include treatments for adenosine deaminase (ADA) deficiency (the severe immune deficiency that requires extreme protection against pathogens and therefore became known as "bubble boy syndrome") (Shaw et al., 2017), adrenoleukodystrophy (ALD) (Eichler et al., 2017), the degenerative brain disease made famous by the movie "Lorenzo's Oil" (1992), and sickle cell disease (Ribeil et al., 2017), the genetic basis of which has been known for more than 60 years.

But another kind of "promising" also has characterized biotech's rise: the incautious, hazardous hawking of premature cures and false solutions. Making these promises helps secure public funding for high-risk investment, fortune seeking, and career building (cf. Joyner et al., 2016). The tragic story of Jesse Gelsinger, what one 1999 *New York Times* article dubbed his "biotech death," brings this dynamic into bold relief (Stolberg, 1999).

The Jesse Gelsinger Story: The Human Cost of Hyped Technology

Eighteen-year old Jesse had suffered from a rare metabolic genetic disease, ornithine transcarbamylase deficiency (OTCD). OTCD prevents the breakdown of ammonia in the body and for children with severe cases it can lead to early death. But Jesse was not sick. His mild version of the disease was controlled by diet and medication. Researchers affiliated with the University of Pennsylvania, keen to test the vaunted promise of manipulating genes to treat disease, told Jesse and his parents that a trial to test a new method for delivering the normal genes to patients' livers was relatively safe. They were told, too, that the results could help babies with more severe OTCD. Jesse also harbored a hope: perhaps someday he would be free of his exacting 32 pills-per-day medication regimen. Primed by promises, Jesse entered the trial generous-hearted and hopeful, and on September 13, 1999 he received an injected delivery of new genes. But a few days later he suffered a massive inflammatory response triggering multiple organ failure and brain death. Ultimately, his devastated parents removed him from life support.

What Jesse and his family had not been told was that two monkeys in the pretrial animal studies had died, that human volunteers had experienced serious adverse reactions, and that the lead researcher, Dr. James M. Wilson, had a serious conflict of interest: his biotech company had a financial stake in the outcome (Obasogie, 2009). Several other companies had invested millions in the technology. Should the gene delivery method prove successful, Dr. Wilson stood to make millions from patenting it. Moreover, the University of Pennsylvania itself was a stockholder in the company associated with the gene protocol. In a subsequent investigation, researchers confessed they had not reported the poor results of earlier research out of their felt need to protect trade secrets.[2] The inquest also concluded that investigators had cut corners: the results of Jesse's physical workup had placed him outside the agreed-upon range for suitable research subjects (Gelsinger and Shamoo, 2008).

Jesse Gelsinger's story underscores aspects still very much in play within biotech culture today: the blurred distinction between researcher and entrepreneur, the rush to imagine ways of monetizing speculative research, the temptation to hype what is being sold to the public – as one bioethicist commented at the time: "Gene Therapy is not yet therapy." Only with the passage of 18 years since Jesse Gelsinger's "biotech death" have effective gene-based treatments and even cures for disabling and fatal diseases emerged (Dunbar et al., 2018).

Part of the delay in bringing these treatments forward was clearly the disrepute brought to the field by the Gelsinger debacle, leading to understandable avoidance by potential investigators and funders (*Nature* editorial, 2016). But the delay also reflects the time it has taken to conduct the basic research required before human applications could be justified. During the two decades since Jesse's death, scientists learned to take a more holistic approach to disease, even those conditions, such as diabetes, heart failure, and colon cancer that have genetic contributions. This has stemmed in part from increased recognition that a given gene does not necessarily function in the same way in different individuals, and even genes previously thought to be uniformly fatal are not inevitably so (Cooper et al., 2013; Hayden, 2016). Other things – stress levels (Uchida et al., 2018), nutritional status (Wallace et al., 2017), even the microbiome (the populations of bacteria inhabiting an individual's digestive tract) (Karkman et al., 2017), can determine whether a variant gene promotes illness or not.

The design of gene therapies, which in some cases appear to be the only means to address a fatal or disabling condition, has vastly improved as science and technology have begun to catch up with the ambitions of developers and commercializers. These therapies fall into several categories, with different levels of hazard. While much more is known than was at the time of Jesse Gelsinger's attempted treatment, in none of these cases have all the risks been fully brought under control.

The Return of Gene Therapy

A variety of gene manipulation methods including CRISPR/Cas9 (discussed in Chapter 6) have allowed accurate, but not fail-safe, targeting of disease-related genes, allowing them to be replaced, or corrected ("edited"), rather than simply supplemented with a normal version, as in Jesse's time. The safest way these methods can be applied is by introducing the corrections into a patient's own cells outside his or her body, observing the cells for sufficient time *in vitro* (i.e., in a culture dish) to confirm that the desired change was made (in subset of the cells, which can be grown in large numbers – "cloned"). These healthily functioning cells can be grafted back into the body. Such an *ex vivo* (outside the body) manipulation was responsible for the dramatic cure by total skin replacement, reported in 2017, of a 7-year-old boy with epidermolysis bullosa, which caused his skin to blister and slough off (Hirsch et al., 2017). Other *ex vivo* genetic manipulations also have shown promise. For example, **T cells** of the immune system have been engineered to fight leukemia (Grady, 2017a).

More challenging, as far as safety is concerned, is targeting a corrective gene to tissues or organs inside the body (referred to as *in vivo* manipulation). This was the plan with Jesse Gelsinger, whose liver was targeted with a gene intended to supplement his malfunctioning one. Depending on the desired change the hazards can vary. In June 2016, biotech company Spark Therapeutics reported that four patients appeared to have been cured of hemophilia after receiving a clotting factor gene that was targeted to the liver (the organ that normally produces and secretes the factor) (Regalado, 2016). This particular procedure has a built-in tolerance for inexactitude, since Factors VIII and IX, which are impaired in the different forms of hemophilia, are blood proteins, which function adequately within a broad range of concentrations.

By contrast, correcting most gene-associated diseases required greater precision. Tay-Sachs disease, for example, is a defect affecting a protein acting inside cells of the brain. Within-cell proteins are subject to much narrower concentration limits than proteins such as Factors VIII and IX which are secreted by cells into the bloodstream. The harmful consequences to the functioning of tissues and solid organs (such as the brain, but not the blood) from defective or dying cells are greater in cases like this, where "misexpression" of the protein could be fatal. This "expression level" problem is far from being solved, which is why *ex vivo* approaches are still the preferred method for solid tissue *ex vivo* gene manipulation. Nonetheless, in some cases, *in vivo* therapies appear to have worked (Bennett et al., 2016; Clark, 2017; Grady, 2017b). The difficulty is reflected in the cost. Spark, the company that developed the hemophilia treatment is now offering an *in vivo* genetic procedure that corrects the retina (a solid tissue) in a rare form of blindness, at $425,000 per eye (Feuerstein and Garde, 2018).

Jesse's death was caused not by misexpression of the protein intended to repair his condition, however, but by the virus used to transmit the genes into his liver cells. The new *in vivo* methods no longer use that viral vector, but a different, disabled one, termed AAV. The safety of delivery of the new genes into the body's tissues is no less important than the accuracy of the manipulation step. So, a class of viruses appearing to bring the delivery problem under control, encouraged the gene therapy community. But experimental trials with nonhuman animals revealed vector-related inflammation problems (Ye et al., 2017). In light of these problems, how will safety be prioritized? Jesse Gelsinger fell victim to a bioscience culture that largely ignored conflicts of interest, cutting corners in pursuit of pushing products to market. Will the present bioscience culture step away from this juggernaut? James Wilson is the physician-scientist who conducted the ill-fated Gelsinger trial. In a stunning move in mid January 2018, Wilson resigned from the board of Solid Biosciences, one of three companies attempting to develop *in vivo* gene correction methods for Duchenne muscular dystrophy (Regalado, 2018). In its press release the company cited Wilson's "emerging concerns about the possible risks of high systemic dosing of AAV" (Walters, 2018). Although Wilson's resignation took place just days before the company's IPO, Solid Biosciences raised more than $125 million dollars. Later that month, Wilson himself revealed that monkeys and pigs either died or experienced disturbing changes. His report sent stock prices of gene therapy companies plunging. (Regalado, 2018). Do Wilson's moves indicate a shift in the cultural winds of biotech development or merely mark how difficult it is to register dissent?

The immediate social aftermath of the Gelsinger family's private tragedy is the tale of a fledgling biotech industry struggling to recapture the glory days of the 1970s genetic revolution's promise. Jesse's story and its social consequences flag a sea-change that had taken place in the biological sciences. The narratives bring sharply into focus how a new era of science entrepreneurialism had begun cresting as the responsible science movement ebbed. Jesse's death occurred almost two decades into a socio-economic bio-research landscape dramatically transformed by two 1980 legal developments. The first was the U.S. Supreme Court decision, *Diamond vs Chakrakarty*. The second was the Bayh-Dole Act. These developments significantly ramped up financial incentives for transferring lab research into commodifiable results.

The Rise of Entrepreneurial Biology: *Diamond v. Chakrabarty* and the Bayh-Dole Act

The U.S. Supreme Court's 1980 ruling in *Chakrabarty* approved the patenting of a genetically engineered oil-eating bacterium. In so doing, it paved the way for the patenting of living organisms and contributed to the private ownership of naturally occurring DNA sequences and cell types. The Court upheld an earlier opinion of the U.S. Court of Customs and Patent Appeals that stated that bacteria are "more akin to inanimate chemical compositions ... [than] to horses and honeybees and raspberries and roses" (Newman, 2003, pp. 439–440). It opened the door to patents on mice, cows, and pigs, some of these mammals containing introduced human genes. Included, too, were naturally occurring human cells as well as nonhuman mammals containing human cells. In the wake of *Chakrabarty* the U.S. Patent Commissioner issued a rule in 1987 stating that the Patent and Trademark Office, "now considers nonnaturally occurring, nonhuman, multicellular living organisms, including animals, to be patentable subject matter" (Newman, 2003, p. 440). The next year Harvard University was granted the first patent for a genetically engineered mammal. The "Oncomouse" was created to develop cancer at a much higher rate than that of its naturally occurring predecessor. In accordance with *Chakrabarty*, the Oncomouse and its offspring (Leder et al., 1986), though alive, were considered "compositions of matter" (one criterion for patentability). University of Michigan Professor of Public Policy, Shobita Parthasarathy succinctly characterized the impact of *Chakrabarty* in *Patent Politics: Life Forms, Markets and the Public Interest in the United States and Europe*:

> The United States has opened its doors wide to life-form patentability, famously allowing patents on "anything under the sun made by man" ... And it has largely dismissed ethical, socio-economic, health, and environmental concerns, characterizing them as distractions in a domain focused on technical questions of novelty, utility, and inventiveness.
>
> (Parthasarathy, 2017, p. 2)

The second legal development, the United States Congress's Bayh-Dole Act, permitted universities and their investigator-employees to privately patent inventions paid for by public funds. Bayh-Dole was a response to the perceived reluctance on the part of industry to invest in federally funded, university research. Traditionally and legally, patent rights to technologies coming from such research remained with the government on behalf of the public. Companies rarely obtained exclusive licenses. The legislation was based on the belief that if universities and their researchers became the patent holders, they would have the freedom to secure venture capital and exclusive corporate licensees. Eventually, Bayh-Doyle's authors hoped, the public would benefit. But instead Bayh-Dole launched an era of an unprecedented, insidious variety of bio-corporate welfare that socialized cost and risk while privatizing profit (Press and Washburn, 2000). When private venture capital shies away from funding basic research that is nowhere near profitable application, the public, via taxes, pays the bills and incurs any losses. In the eventuality of successful discoveries, however, profit flows to the private patent holder. And, as we will see later in this book, this feeding at the public trough is facilitated greatly by inflated promises and deception, diverting interest and investment from less flashy but possibly more useful inventions.

Although directed to all federally financed science and engineering-based technologies and not specifically to the biological sciences, Bayh-Dole particularly had a powerful effect on the biotech revolution. Like the historic enclosures that removed arable lands from general use in England centuries ago, tweaked bits of DNA began fencing off the biological commons. Patents issued on naturally occurring DNA sequences converted a common resource into one with exclusive private beneficiaries. The new biology of the 1980s and 1990s came to bear a commercial stamp unlike any bio-science research coming before it. Ultimately, in the new highly commercially charged, bio-private propertied post Bayh-Dole world, the cultural impact of having learned to recombine DNA had an ironic effect. It shriveled the shared knowledge-for-knowledge sake curiosity that first fueled its discovery. The informal give-and-take characterizing earlier biological research declined sharply. Instead, conflicts of interest previously unknown to cell and developmental biology and related fields grew paramount.

Bayh-Dole also reoriented major universities to seeing their intellectual property portfolios as means for vastly improving their financial bottom lines. Although technology transfer, moving basic research from the ivory tower to broader use, had long been a university function, the new era of entrepreneurial bioscience was troubling to many both within and outside the circles of professional science and corridors of higher education. "The new-found concern with technology transfer is disturbing," wrote Harvard President Derek Bok in 1982, "not only because it could alter the practice of science in the university but also because it threatens the central values and ideals of academic research." "It is one thing," he judged, "to consult for a few hundred dollars a day or to write a textbook in the hope of receiving a few thousand dollars a year. It is quite another matter to think of becoming a millionaire by exploiting a commercially attractive discovery."

By the mid 1980s, historian Charles Weiner had identified a "credibility gap." "The dual roles that many biologists play," he analyzed, "have begun to impair their credibility on matters of public concern . . . The conflicts of interest created by the commercialization of academic biology could erode public trust." For Weiner, this erosion came at a crucial point in time when the expert advice of biologists was "urgently needed on questions of public policy." He cast the net of questioning expansively, implicitly resurrecting the broad roster of concerns remaindered so casually by bioengineers at Asilomar:

> These questions range from deciding priorities in the application of biotechnology, to whether genetically engineered organisms should be deliberately released into the environment, to the ethical implications of human gene therapy, to the use of biotechnology in developing biological weapons.
>
> (Weiner, 1986, p. 43)

Biologists had an important role to play in these deliberations. But, Weiner feared, "their commercial roles may well affect their perceptions and limit their effectiveness as credible public advisors" (Weiner, 1986).

By the opening decade of the twenty-first century, a few writers and journalists had begun recognizing the seismic shifts in the doing of science triggered by the 1980 legal developments. In 2000, Eyal Press and Jennifer Washburn identified a new "academic-industrial complex," for *Atlantic Monthly* readers in their essay, "The Kept University." They placed the Bayh-Dole Act within the context of declining U.S. productivity and overseas competition, and the increasing consolidation between university research and commercial enterprise. "In an age when ideas are central to the economy, universities will inevitably play a role in fostering growth," they acknowledged. But Bayh-Dole's sullying effect on academia disturbed them: "[S]hould we allow commercial forces to determine the university's educational mission and academic ideals?," they asked pointedly (Press and Washburn, 2000).

In 2001, science journalist Tom Abate described these developments vividly:

> When we spliced the profit gene into academic culture, we created a new organism – the recombinant university. We reprogrammed the incentives that guide science. The rule in academe used to be "publish or perish." Now bioscientists have an alternative – "patent and profit."

He offered the example of University of California scientists, who won a patent for a technique that put "genes into pills." Instead of sharing and publishing their results as would be expected traditionally, they raised money to start a company to "try and make gene pills as medicines." "[I]f academic science has changed," he asked, "why should we trust its self-policing mechanisms?" (Abate, 2001).

In 2002, science reporter Neil Munro gave the players operating within the controversial new terrain a name: "scientist-entrepreneurs." "By keeping one foot in business and one in the university, these scientist-businessmen," explained Munro,

> get the best from both worlds. As academics, they get plaudits from their peers, the approval of colleagues serving on federal grant-review boards, and early access to hot new discoveries – not to mention federally subsidized laboratory space and an endless supply of underpaid graduate students eager to help develop the next cutting-edge idea. As entrepreneurs, they get access to wealth and the investment capital needed to launch a first, second, or third company.

For Munro, the Janus-face of the scientist-entrepreneur made the socio-political complexities of science and medicine especially perplexing to sort through. It still does. Is it possible to find an objective arbiter of ethical issues when those most able to explain the research in question are also those who stand to profit lavishly from its applications? "These supposedly objective scientists," explained Munro, "have business interests that overlap with their scientific views." Often they expect to market resulting procedures through biotech companies that they themselves start up. "The problem," Munro felt, "lies with the press, which almost never informs its readers that these supposedly disinterested scientists have great financial stakes in the debate" (Munro, 2002). In 2013, Jeremy Gruber, former President of the Council for Responsible Genetics, would zero-in on a key import of these new academic-commercial connections:

> Academic researchers form much of the initial staff of biotech start-up companies, and as these companies grow, they form lucrative partnerships with universities that include everything from collaborative research to consulting relationships. Many such researchers sit on the advisory boards of biotechnology companies. As a result, academic researchers are often the biggest industry boosters.
> (Krimsky and Gruber, 2013, p. 277)

The observation is one that science journalists today seem to forget to remember.

Commerce Eclipsing Criticism

Indeed, those journalists offering introspective and critical comment at the turn of the millennium, though few in number, evidenced a concern almost wholly undetectable in journalism today, barely over a decade later. At present, troubling ramifications of biotech commerce appear to be invisible to science chroniclers peering out through the new normal. Often enough, scientists are identified solely by university affiliation, their role in founding biotech companies or serving on bio-industry boards going unmentioned or, if referenced at all, slipped in as uncritiqued afterthought. But the social and ethical effects of commercialized bioscience, when first recognized, were jarring. Bayh-Dole inspired entrepreneurialism seemed founded on twin piers of greed and conflicts of interest. And compounding the problem of potentially corrupting influences was, and remains, the failure of bioentrepreneurs to recognize them as such. Tufts University Professor Sheldon Krimsky analyzed conflicts of interest in science in his 2003 book, *Science in the Private Interest: Has the Lure of Profits Corrupted Biomedical Research?* Krimsky explains that typical scientists are

> incredulous that any financial interest they might have connected to their research would affect the way they do science ... It is widely accepted among members of the scientific community that the "state of mind" of the scientist is not prone to the same influences that are known to corrupt the behavior of public officials.
>
> (pp. 129–130)

In their 2012 account as "participant-observers" embedded at the Synthetic Biology Research Center (Synberc, see Chapter 5), medical anthropologist Paul Rabinow and ethicist Gaymon Bennett offer a sharp characterization of their bioscientist and engineer colleagues that serves to challenge the scientist's self-perceived ethically incorruptible state of mind that Krimsky identified: "All of the ... players involved are unashamedly and unselfconsciously committed to prospering personally, institutionally, or nationally – most proceed as though all three are synergistically connected" (Rabinow and Bennett, 2012, pp. 7–8). When the authors sought to work collaboratively with synthetic biologists on the ethics and social implications of their work, they found that, "we were often ignored by the senior members or met with overt hostility from younger scientists who saw our interventions as an encroachment on their time and career goals" (Rabinow and Bennett, 2012, p. 29). Do financial and career influences, including those unrecognized by scientists as such, make clear-sighted consideration of the responsible science generation's roster of concerns possible? This question is of particular importance since, rising in tandem with the promise of genetic cures, is the potential to manipulate human biology to practical and commercial ends, including creation of part-human animals, including some with potentially human-like consciousness (see Chapters 2 and 6).

The decades following the advent of gene splicing, and the financial and social transfigurations that rose in its wake, saw the rise of other biotechnologies with capabilities of realizing the worst fears of postwar critics in the "responsible scientist" tradition. Much of this has been enabled by a generally benign procedure, *in vitro fertilization* (IVF). Developed first by animal breeders, IVF enabled the fertilization of egg and sperm outside the human body. Following the 1978 birth in England of Louise Brown, the first "test-tube baby," millions of children have been born world-wide using this technique, most often to parents who couldn't have conceived otherwise. This massive intervention into the reproductive process is acknowledged to be an experiment, the health effects of which on IVF-produced individuals will not be known for several generations (Knapton, 2016). The desire to have a child led many to take the risk. But beyond this original motivation for using IVF, the possibility of manipulating embryos for eventual implantation dramatically sparked imaginations, provoking plans ranging from the elimination of genetic diseases to the creation of "designer babies" possessing enhanced features and abilities. How would this spectrum of possibilities play out in an ethical arena denatured by bio-enclosures occasioned by Bayh-Dole-*Chakrabarty*?

One thing became clear: when the capacity to implement genetic technologies that Tiselius warned about in 1961 approached fruition, commercial incentives would loom large among the forces shaping science. These incentives could result in cheaper more effective applications. But they could also provoke deception, false promises, and injurious practices. Today, social and ethical aspects of science and technology often is overseen by the new profession of "bioethics," which rose in tandem with biotechnology (Stevens, 2003). Does such generally sympathetic handling amplify broad social awareness, including the effects of financial and careerist influences? Or does it dampen controversy, channeling it away from the bright light of unprimed public scrutiny?

What Is to Come (in this Book?; in the Future?)

As a field deeply involved with the anxieties associated with actual and prospective illness, biotechnology is particularly prone to overheated promises and financial opportunism. This has been seen in some extreme cases that have come to the public's attention. The principals of ImClone indulged in insider trading when FDA evaluations of their highly touted anti-cancer therapeutics did not turn out as expected, leading to prison time for them and one of their high-profile investors, the homemaking guru Martha Stewart (Ulick, 2003). More recently, Theranos, a company with more than $400 million of venture capital funding and an eventual valuation of more than

$9 billion, collapsed spectacularly when the diagnostic testing instrumentation marketed by its charismatic but deceptive CEO Elizabeth Holmes turned out to be ineffective (*Economist*, 2018).

The sector of the biotechnology industry that is the main subject of this book relates to the intersection of cell and tissue research with the objective of modifying the cells and organs of the human body, and ultimately, if allowed to run its current course, remaking it in part or whole. This is a speculative and poorly understood field of research, more removed from the classical concerns of medicine than the sectors involved in drug development and diagnostics. It has the potential to change civilization, and as we will see, is subject to a unique array of promises and deceptive marketing practices, some, in the form of "transhumanism," verging on science fiction. The following chapters of *Biotech Juggernaut* reflect on the cultural politics of science in the age of bioentrepreneurialism as they relate to human genetic engineering.

In Chapter 2, "The Dawn of GM Humans" we describe the emergence of some of the technologies anticipated by postwar scientists, their promise, their limitations, and their challenges, both technical and social. We relate the historic and scientific significance of the 2000 sequencing of the human **genome**, including its effect of culturally reinforcing the fallacy of genetic determinism (the erroneous "one-gene, one-trait" correlation, but also its more sophisticated but equally misleading replacements based on genetic circuits and networks). We then consider three emerging biotechnologies with serious implications for genetically modifying humans: **cloning** (including the prospect, realized in nonhuman animals, of full-term cloning), **embryonic stem cells**, and embryo gene modification. The chapter outlines how, through implementation of these biotechnologies, we are approaching an era of "organoids" based on human embryos that while plausibly serving human needs, will inevitably come to blur the boundary between humans and industrial products. The role of bioentrepreneurs in redefining scientific terminology to disguise the controversial aspects of the biotechnologies they pursue is discussed.

Chapters 3 and 4, "California Cloning: The Campaign" and, "California Cloning: The Aftermath," offer an in-depth examination of the biotech juggernaut at work, relating how political dynamics surrounding some of the technologies discussed in Chapter 2 affected events centered in California, a roiling epicenter of biotech financing, research, and development. We recount the passage of the 2004 California Stem Cell Research and Cures Act, which created the powerful and highly controversial publicly funded **California Institute of Regenerative Medicine** (CIRM). This case study offers an opportunity to examine troubling practices adopted by bioentrepreneurs in an overzealous promotion of the contentious initiative, "Proposition 71" and in the administration of the initiative once passed. These practices include: redefining terms to avoid public recognition of contentious aspects (e.g., Prop 71's prioritization of cloning technology, that technology's need of women's eggs, and the health risks to women of acquiring those eggs), camouflaging controversies behind scientific jargon, hyping the promise and possibilities for patented applications, bringing legal action to silence critics, and concealing marketplace conflicts of interest by cloaking corporate titles under the feigned neutrality of academic credentials.

Chapter 5, "Synthetic Biology: Extreme Genetic Engineering," is the only portion of the book to describe more general questions of bioengineering, but primarily because techniques such as **CRISPR/Cas9** gene manipulation being developed in the nonhuman and even non-animal realm are already being used in attempts to genetically engineer prospective people. The chapter covers developments in the emerging science and technology platform, "**synthetic biology**."

By the opening decade of the twenty-first century, we had entered a new era of biotechnological transformation. Synthetic biology entrepreneurs seek to create life using computer-designed or artificial DNA. As one synthetic biologist characterized it, genetically engineered bacteria in an industrial vat can now create anything traditionally harvested from a plant. Such hijacking of microbial processes is resulting in vast fortunes for many biocorporations in the industrial north. But for traditional guardians of plant-based economies, chiefly farming and peasant societies in the global south, synthetic biology as practiced destroys livelihoods and communities. And, since the food required to feed these putatively synthetic organisms is sugar, vast expanses of the world's forested areas are being transformed from pristine regions to commercial sugar cane production. For consumers, too, synthetic biology has been transformational, although in ways unknown to most. Because synthetic biology is part of biotech's seamless blend of research and application, it affects production rapidly, transfiguring our world even while its processes are still being identified. Already, consumption of products of synthetic biology, from fuels and cleansers to foods, fragrances and cosmetics, is both part of the new field's trial-and-error experimental process and its industrial platform. In the blink of a historic eye, products of extreme genetic technologies have stealthily infiltrated daily life, redirecting evolution, creating new classes of producers, devastating traditional economies, and silently altering the purchasing behaviors of unsuspecting consumers. The tacit and sometimes explicit rejection of valuation of natural biological systems refined through eons of evolution over synthetic ones modified for narrow purposes is underway. It carries ominous implications for proposed uses of these new methods in human reproduction discussed in the chapter that follows.

Chapter 5 also engages an ongoing inquiry of whether suitably extreme regulatory caution has accompanied this "extreme genetic engineering." What are the safety and ecological implications of the release of novel, self-replicating organisms? What do some of the field's practitioners intend for the future of the human species and how are they selling this vision? It also relates the efforts of the Lawrence Berkeley National Laboratory and the University of California to create what would have constituted the world's largest publicly funded synthetic biology lab.

Chapter 6 offers an update on human genetic engineering, analyzing the "three-parent IVF" technology and the gene "editing" CRISPR/Cas9 technology. "The Road to Gattaca," parses the language used to describe these technologies and recognizes the use of two strategies observed throughout *Biotech Juggernaut* to be persistently employed by stakeholders interested in propelling controversial technologies into broad acceptance: first, using terminology that appears neutral but which actually masks controversial aspects of the methodologies in question and second, framing all discussion in terms of guarantees of cures. These strategies pay little attention either to the potential scientific barriers to obtaining such cures or, even if cures should be found, to very troubling developments that accompany such agendas, e.g., the creation of quasi-human beings and the promulgation of techno-eugenics. Specifically, we discuss the **chimera** technique developed in the 1980s, whereby embryos of different animal species were blended together to yield new kinds of composite organisms. When co-author Stuart Newman used the patent application process to bring to the public's attention the implications of this technique for blurring the human–nonhuman barrier, his concerns were largely dismissed, but the predicted uses have come to pass in the intervening decades. We will also revisit the crucial distinction between **somatic cell** modification and **germline** genetic modification with respect to the quest for cures.

Finally, in Chapter 7, we reflect on possibilities for robust advocacy on behalf of preserving a human future. This chapter reflects some of our motivation for undertaking this project. *Biotech Juggernaut* seeks to put historical and scientific explanation in the service of widening understanding of and engagement with extraordinarily significant issues; and, hopefully, to liberate these issues from the sheltered confines of elite, enabling deliberation.

Notes

1 The experience of this book's co-author, Stuart Newman, as a member of Science for the People (SftP) and a co-founder of Council for Responsible Genetics (CRG) offers insight into their role as torch-bearers of the earlier responsible science movement. SftP and CRG were both avowedly inspired by the *Bulletin of Atomic Scientists* (BAS) and the anti-nuclear weapons and anti-war movement, but were more comprehensive than the BAS in the issues they took on, and, in the case of SftP, which arose in the social environment of the '60s and early '70s, more explicitly radical in its vision for the political change required to harness technology for the public good. SftP members were also involved in developing critiques of the ideology of basic research itself, in both the physical and biological (e.g., sociobiology,

agriculture) sciences, seeing notions of acceptable practical applications inextricably tied to theoretical considerations. While CRG confined itself to the biological sciences, the scope of these had expanded enormously by the early '80s when it was formed. Although some of the founding members of CRG were veterans of StfP, the later organization was from the start less ideologically oriented than the earlier one, and also included social scientists, labor activists, and policy analysts among its board members. Other scientists with early connections to StfP followed different routes, relinquishing a critical stance toward science and technology, often partaking of academic and commercial opportunities characteristic of the post-Bayh-Dole and *Chakrabarty* decision, Reagan and Thatcher period. For a recent account of the history of SftP see, *Science for the People: Documents from America's Movement of Radical Scientist* by Schmalzer, Chard, and Botelho (University of Massachusetts Press, January, 2019.) An account of the history of CRG, outside the scope of *Biotech Juggernaut*, awaits historical characterization.

2 While the Recombinant DNA Advisory Committee (RAC) that approved the U. Penn protocol was bound to respect trade secrets in their closed reviews, the investigators were required to disclose all details to the RAC and FDA. This undermined the investigators' claims.

Sources Consulted for Chapter 1

Abate, Tom, "Scientists' 'Publish or Perish' Credo Now 'Patent and Profit' / 'Recombinant U.' Phenomenon Alters Academic Culture," SF Gate, August 13, 2001: www.sfgate.com/business/article/Scientists-publish-or-perish-credo-now-patent-2891077.php

Bennett, J., J. Wellman, K.A. Marshall, S. McCague, M. Ashtari, J. DiStefano-Pappas, O.U. Elci, D.C. Chung, J. Sun, J.F. Wright, D.R. Cross, P. Aravand, L.L. Cyckowski, J.L. Bennicelli, F. Mingozzi, A. Auricchio, E.A. Pierce, J. Ruggiero, B.P. Leroy, F. Simonelli, K.A. High, and A.M. Maguire, "Safety and Durability of Effect of Contralateral-Eye Administration of AAV2 Gene Therapy in Patients with Childhood-Onset Blindness Caused by RPE65 Mutations: A Follow-On Phase 1 Trial," *Lancet* 388 (2016): 661–672.

Billings, P.R., R. Hubbard, and S.A. Newman, "Human Germline Gene Modification: A Dissent," *Lancet* 353 (1999): 1873–1875.

BIO website: www.bio.org/articles/what-biotechnology

Bok, Derek. *Beyond the Ivory Tower: Social Responsibilities of the Modern University*, Cambridge, MA, Harvard University Press, 1982.

Buck v. Bell, 274 U.S. 200 (1927) https://caselaw.findlaw.com/us-supreme-court/274/200.html

Bud, Robert, *The Uses of Life: A History of Biotechnology*, Cambridge, UK, Cambridge University Press, 1994.

Church, George, "The Augmented Human Being." A Conversation with George Church, March 30, 2016. *The Edge*: www.edge.org/conversation/george_church-the-augmented-human-being

Clark, Toni, "Gene Therapy for Blindness Appears Initially Effective, Says U.S. FDA," *Scientific American*, October 10, 2017: www.scientificamerican.com/article/gene-therapy-for-blindness-appears-initially-effective-says-u-s-fda/

Comfort, Nathaniel, *The Science of Human Perfection: How Genes Became the Heart of American Medicine*, New Haven, CT: Yale University Press, January, 2014.

Cooper, D.N., M. Krawczak, C. Polychronakos, C. Tyler-Smith, and H. Kehrer-Sawatzki, "Where Genotype Is Not Predictive of Phenotype: Towards an Understanding of the Molecular Basis of Reduced Penetrance in Human Inherited Disease," *Hum Genet* 132 (2013): 1077–1130.

Diamond vs. Chakrabarty, 447 U.S. 303 (1980).

Economist, "The Rise and Fall of Elizabeth Holmes, Silicon Valley's Startup Queen," May 31, 2018: www.economist.com/books-and-arts/2018/06/02/the-rise-and-fall-of-elizabeth-holmes-silicon-valleys-startup-queen

Dunbar, C.E., K.A. High, J.K. Joung, D.B. Kohn, K. Ozawa, and M. Sadelain, "Gene Therapy Comes of Age," *Science* 359 (January 12, 2018).

Eichler, F., C. Duncan, P.L. Musolino, P.J. Orchard, S. De Oliveira, A.J. Thrasher, M. Armant, C. Dansereau, T.C. Lund, W.P. Miller, G.V. Raymond, R. Sankar, A.J. Shah, C. Sevin, H.B. Gaspar, P. Gissen, H. Amartino, D. Bratkovic, N.J.C. Smith, A.M. Paker, E. Shamir, T. O'Meara, D. Davidson, P. Aubourg, and D.A. Williams, "Hematopoietic Stem-Cell Gene Therapy for Cerebral Adrenoleukodystrophy," *N Engl J Med* 377 (2017): 1630–1638.

Feurestein, A., and D. Garde, "Spark Prices Its Gene Therapy as Most Expensive U.S. Medicine – but with Plans to Ease Cost Concerns," *STAT*, January 3, 2018: www.statnews.com/2018/01/03/spark-gene-therapy-price/

Gelsinger, Paul, and Adil Shamoo, "Eight Years after Jesse's Death, Are Human Research Subjects Any Safer?" *Hastings Center Report*, April 4, 2008.

Grady, Denise, "F.D.A. Approves Second Gene-Altering Treatment for Cancer," *New York Times*, October 18, 2017a: www.nytimes.com/2017/10/18/health/immunotherapy-cancer-kite.html

Grady, Denise, "Gene Therapy Creates Replacement Skin to Save a Dying Boy," *New York Times*, November 8, 2017b: www.nytimes.com/2017/11/08/health/gene-therapy-skin-graft.html

Grogan, Suzie, Shell Shocked Britain: The First World War's Legacy for Britain's Mental Health, Barnsley, South Yorkshire: Pen & Sword History, 168pp.

Hall, Leslie, "Illustrations from the Wellcome Institute Library, the Eugenics Society Archives in the Contemporary Medical Archives Centre," *Medical History* 34 (1990): 327–333: www.cambridge.org/core/services/aop-cambridge-core/content/view/S0025727300052467

Hayden, E., "Seeing Deadly Mutations in a New Light," *Nature* (2016): 154–157.

Hayes, Richard, "In the Pipeline: Genetically Modified Humans?" *Multinational Monitor*, January/February, 2000: www.multinationalmonitor.org/mm2000/012000/hayes.html

Hirsch, T., T. Rothoeft, N. Teig, J.W. Bauer, G. Pellegrini, L. De Rosa, D. Scaglione, J. Reichelt, A. Klausegger, D. Kneisz, O. Romano, A. Secone Seconetti, R. Contin, E. Enzo, I. Jurman, S. Carulli, F. Jacobsen, T. Luecke, M. Lehnhardt, M. Fischer, M. Kueckelhaus, D. Quaglino, M. Morgante, S. Bicciato, S. Bondanza, S., and M. De Luca, "Regeneration of the Entire Human Epidermis Using Transgenic Stem Cells," *Nature* 551 (2017): 327–332.

Joyner, Michael J., Nigel Paneth, and John P.A Ioannidis, "What Happens When Underperforming Big Ideas in Research Become Entrenched?" *The Journal of the American Medical Association*, July 28, 2016: https://jama.jamanetwork.com/article.aspx?articleid=2541515

Karkman, A., J. Lehtimaki, and L. Ruokolainen, "The Ecology of Human Microbiota: Dynamics and Diversity in Health and Disease," *Ann N Y Acad Sci* 1399 (2017): 78–92.

Kevles, Daniel J., In the Name of Eugenics: Genetics and the Uses of Human Heredity, New York: Knopf, 1985.

Knapton, Sarah, "Test Tube Babies Could Die Sooner," *The Telegraph*, February 16, 2016: www.telegraph.co.uk/news/science/science-news/12158072/Test-tube-babies-could-die-sooner.html

Kolata, Gina, "Building a Better Human with Science? The Public Says, No Thanks," *New York Times*, July 26, 2016: www.nytimes.com/2016/07/27/upshot/building-a-better-human-with-science-the-public-says-no-thanks.html?_r=0

Kreuger, Annette, "And So It Grows . . . $136 Billion in Pharmaceutical-Biotech Facility Investments Seen around the Globe," *Industrial Info Resources*, February 9, 2018: www.interphex.com/RNA/RNA_Interphex_V2/2018/_docs/INTERPHEX-NY-2018-Technical-White-Paper-Industrial-Info-Resources.pdf?v=636549271373662601

Krimsky, Sheldon, *Genetic Alchemy: The Social History of the Recombinant DNA Controversy*, Cambridge, MA: MIT Press, 1985.

Krimsky, Sheldon. *Science in the Private Interest: Has the Lure of Profits Corrupted Bioemedical Research?*, Lanham, MD: Rowman & Littlefield, 2003.

Krimsky, Sheldom, and Jeremy Gruber, *Genetic Explanations: Sense and Nonsense*, Cambridge, MA: Harvard University Press, 2013.

Leder, A., P.K. Pattengale, A. Kuo, T.A. Stewart, and P. Leder, "Consequences of Widespread Deregulation of the C-Myc Gene in Transgenic Mice: Multiple Neoplasms and Normal Development," *Cell* 45 (1986): 485–495.

Lederberg, Joshua, "Molecular Biology, Eugenics, and Euphenics," *Nature* 198(4879) (May, 1963): 428–429.

Lederberg, Joshua, "Experimental Genetics and Human Evolution," *The American Naturalist* 100(915) (1966): 519–531.

Lippman, A., P. Bereano, P. Billings, C. Gracey, M.S. Henifin, R. Hubbard, S. Krimsky, R.C. Lewontin, K. Messing, S. Newman, J. Norsigian, M. Saxton, D. Stabinsky, and N.L. Wilker, "Position Paper on Human Germ Line Manipulation Presented by Council for Responsible Genetics, Human Genetics Committee Fall, 1992," *Human Gene Therapy* 4(1) (1993): 35–37.

Lorenzo's Oil IMDb, 1992: www.imdb.com/title/tt0104756

Munro, Neil, "Dr. Who? Scientists Are Treated as Objective Arbiters in the Cloning Debate. But Most Have Serious Skin in the Game," *Washington Monthly*, November 2002: www.washingtonmonthly.com/features/2001/0211.munro.html

Newman, Stuart A., "Averting the Clone Age: Prospects and Perils of Human Developmental Manipulation," *Journal of Contemporary Health Law and Policy* 19(1) (2003): 431–463.

Nature, editorial, "Gene Therapy Trials Must Proceed with Caution," June 28, 2016: www.nature.com/news/gene-therapy-trials-must-proceed-with-caution-1.20186?WT.mc_id=TWT_NatureNews

Obasogie, Osagie K., "Ten Years Later: Jesse Gelsinger's Death and Human Subjects Protection," *The Hastings Center Bioethics Forum Essay*, October 22, 2009: www.thehastingscenter.org/ten-years-later-jesse-gelsingers-death-and-human-subjects-protection/

Parthasarathy, Shobita, *Patent Politics: Life Forms, Markets and the Public Interest in the United States and Europe*, Chicago, IL: University of Chicago Press, 2017.

PitchBook, "VC Investment in Biotech Blasts through $10B Barrier in 2017," 2017: https://pitchbook.com/news/articles/vc-investment-in-biotech-blasts-through-10b-barrier-in-2017

Press, Eyal, and Jennifer Washburn, "The Kept University," *The Atlantic*, March 2000: www.theatlantic.com/magazine/archive/2000/03/the-kept-university/306629/

Rabinow, Paul, and Gaymon Bennett, *Designing Human Practices: An Experiment with Synthetic Biology*, Chicago, IL: University of Chicago Press, 2012.

Regalado, Antonio, "Gene Therapy Is Curing Hemophilia," *MIT Technology Review*, June 11, 2016: www.technologyreview.com/s/601651/gene-therapy-is-curing-hemophilia/?utm_campaign=content-distribution&utm_source=dlvr.it&utm_medium=twitter

Regalado, Antonio, "The Doctor Responsible for Gene Therapy's Greatest Setback Is Sounding a New Alarm," *MIT Technology Review*, January 31, 2018: www.technologyreview.com/s/610141/the-doctor-responsible-for-gene-therapys-greatest-setback-is-sounding-a-new-alarm/

Ribeil, J.A., S. Hacein-Bey-Abina, E. Payen, A. Magnani, M. Semeraro, E. Magrin, L. Caccavelli, B. Neven, P. Bourget, W. El Nemer, P. Bartolucci, L. Weber, H. Puy, J.F. Meritet, D. Grevent, Y. Beuzard, S. Chretien, T. Lefebvre, R.W. Ross, O. Negre, G. Veres, L. Sandler, S. Soni, M. de Montalembert, S. Blanche, P. Leboulch, and M. Cavazzana, "Gene Therapy in a Patient with Sickle Cell Disease," *N Engl J Med* 376 (2017): 848–855.

Shaw, K.L., E. Garabedian, S. Mishra, P. Barman, A. Davila, D. Carbonaro, S. Shupien, C. Silvin, S. Geiger, B. Nowicki, E.M. Smogorzewska, B. Brown, X. Wang, S. de Oliveira, Y. Choi, A. Ikeda, D. Terrazas, P.Y. Fu, A. Yu, B.C. Fernandez, A.R. Cooper, B. Engel, G. Podsakoff, A. Balamurugan, S. Anderson, L. Muul, G.J. Jagadeesh, N. Kapoor, J. Tse, T.B. Moore, K. Purdy, R. Rishi, K. Mohan, S. Skoda-Smith, D. Buchbinder, R.S. Abraham, A. Scharenberg, O.O. Yang, K. Cornetta, D. Gjertson, M. Hershfield, R. Sokolic, F. Candotti, and D.B. Kohn, "Clinical Efficacy of Gene-Modified Stem Cells in Adenosine Deaminase-Deficient Immunodeficiency," *J Clin Invest* 127 (2017): 1689–1699.

Stern, A., *Eugenic Nation: Faults and Frontiers of Better Breeding in Modern America*, Oakland, CA: University of California Press, 2016.

Stevens, M.L. Tina, *Bioethics in America: Origins and Cultural Politics*, Baltimore, MD: Johns Hopkins University Press, paperback edition, 2003.

Stolberg, Cheryl Gay, "The Biotech Death of Jesse Gelsinger," *The New York Times*, 1999: www.nytimes.com/1999/11/28/magazine/the-biotech-death-of-jesse-gelsinger.html? src=pm&pagewanted=1

Uchida, S., H. Yamagata, T. Seki, and Y. Watanabe, "Epigenetic Mechanisms of Major Depression: Targeting Neuronal Plasticity," *Psychiatry Clin Neurosci* 72(4) (2018): 212–227.

Ulick, Jake, "Martha Indicted, Resigns," *CNN/Money* June 4, 2003: http://money.cnn.com/2003/06/04/news/martha_indict/index.htm

Wallace, R.G., L.C. Twomey, M.A. Custaud, J.D. Turner, N. Moyna, P.M. Cummins, and R.P. Murphy, "The Role of Epigenetics in Cardiovascular Health and Ageing: A Focus on Physical Activity and Nutrition," *Mech Ageing Dev* 174 (2018): 76–85.

Walters, Jenny, "James Wilson Resigns from Solid Biosciences' Board," *Biocentury*, January 17, 2018: www.biocentury.com/bc-extra/company-news/2018-01-17/james-wilson-resigns-solid-biosciences-board

Washburn, Jennifer, *University Inc. The Corporate Corruption of Higher Education* New York: Basic Books, paperback edition, 2006, Hardback, 2005.

Weiner, Charles, "Universities, Professors, and Patents: A Continuing Controversy," *Technology Review* (February/March, 1986): 33–43.

Weiner, Charles. "Social Responsibility in Genetic Engineering: Historical Perspectives," Anders Nordgren, ed., *Gene Therapy and Ethics, Studies in Bioethics and Research Ethics* 4, Uppsala: Acta Universitatis Upsaliensis, 1999, 51–64.

Wright, Susan, "Molecular Biology or Molecular Politics? The Production of Scientific Consensus on the Hazards of Recombinant DNA Technology," *Social Studies of Science* 16(4) (November, 1986): 593–620.

Wright, Susan, *Molecular Politics: Developing American and British Regulatory Policy for Genetic Engineering, 1972–1982*, Chicago, IL: University of Chicago Press, 1994.

Ye, G.J., E. Budzynski, P. Sonnentag, T.M. Nork, P.E. Miller, A.K. Sharma, J.N. Ver Hoeve, L.M. Smith, T. Arndt, R. Calcedo, C. Gaskin, P.M. Robinson, D.R. Knop, W.W. Hauswirth, and J.D. Chulay, "Safety and Biodistribution Evaluation in Cynomolgus Macaques of rAAV2tYF-PR1.7-hCNGB3, a Recombinant AAV Vector for Treatment of Achromatopsia," *Hum Gene Ther Clin Dev* 27(1) (2016): 37–48.

Zilinskas, Raymond, "The Promise and Perils of Synthetic Biology," *The New Atlantis* 12 (Spring 2006): 25–45.

2
The Dawn of GM Humans

As we move into an era of advanced germinal choice, children conceived with these technologies will necessarily intermingle with those with more haphazard beginnings. But how they will relate to one another in the long run is no more clear than whether a gulf will ultimately widen between them, partitioning humanity into the enhanced and the unenhanced.

(Gregory Stock, Biophysicist, Co-founder, Signum Biosciences, Inc. from, *Redesigning Humans: Our Inevitable Genetic Future*, 2002)

If the accumulation of genetic knowledge and advances in genetic enhancement technology continue at the present rate, then by the end of the third millennium, the GenRich class and the Natural class will become the GenRich-humans and the Natural-humans – entirely separate species with no ability to cross-breed, and with as much romantic interest in each other as a current human would have for a chimpanzee . . . I leave it to philosophers and bioethicists to figure out how . . . ethical dilemmas might be resolved.

(Lee Silver, Biologist, Co-founder, GenePeeks, Inc. from, *Remaking Eden: How Genetic Engineering and Cloning Will Transform the American Family*, 1997)

Biochemist Marshall Nirenberg won the Nobel Prize in 1968 for deciphering the genetic code. A scant year earlier, however, the lauded researcher struck a cautionary tone about scientific investigation. "When man becomes capable of instructing his own cells," he warned in *Science* magazine, "he must refrain from doing so until he has sufficient wisdom to use this knowledge for the benefit of mankind" (cited in Weiner, 1999, p. 54). But the biomolecular revolution to which Nirenberg contributed was moving even faster than imagined and outpaced admonitions. By the closing decades of the twentieth century, the biosciences were enmeshed in some of the very ethical controversies that Nirenberg and others found so troubling. The 1970s to the 1990s saw the dawn of an intensified inquisitiveness about manipulating human embryos. Researchers were uncovering information, and in some cases formulating agendas, about human genetic engineering.

Biotechnologies anticipated by scientists-critics during the waning of the postwar responsible science movement were coming of age. These technologies rose in an era transformed by the post-1980 *Chakrabarty*–Bayh-Dole financial revolution (see Chapter 1). The overhaul of patent application and ownership laws rapidly altered the doing of science and intensified troubling aspects of the biotechnologies themselves (Parthasarathy, 2017). In significant ways, commercial incentives and conflicts of interest came to direct both the science and its promotion in civil society. Promoters employed a number of strategies to edge around ethical hurdles, especially those concerning the creation of clones and chimeras, the derivation of embryonic stem cells, and the genetic modification of embryos (see pp. 32ff). Some controversies became so politically charged that they jeopardized continuing support from the national government, the traditional source of science research funding.

Feeling financially forsaken by the federal government, bio-entrepreneurs made the unprecedented move of turning to the states with their hands out. They hoped that California, in particular, would yield a gold rush for the new millennium. And it did. By 2004 that state, despite its own colossal debt crisis, pledged $3 billion (becoming $6 billion upon repayment of the issued bond), to create and finance the California Institute of Regenerative Medicine, CIRM. Seduced by a plea and in exchange for a pledge, Californians willingly shouldered this astonishing burden. Voters encumbered the state's budget believing the promise that if only funding could be found, cures for every manner of disease would be available imminently. They did this despite the fact that venture capitalists would not. They did this unable to discern that CIRM's mandate would encourage the risky harvesting of thousands of women's eggs needed as raw material for research that would further human cloning, a biotechnology the vast majority of polled Americans strongly opposed (Beeson and Stevens, 2004).

Nirenberg's appeal for wisdom to guide applications of the research he and his peers were undertaking reflected a confidence about where that wisdom would be found: "I state this problem well in advance of the need to resolve it," he explained, "because decisions concerning the application of this knowledge must ultimately be made by society and only an informed society can make such a decision wisely" (cited in Weiner, 1999, p. 54). CIRM, described in Chapters 3 and 4, is a manifest thwarting of this buoyant expectation. The state ballot proposition supporting its funding, "Proposition 71," had been put to a vote after a well-funded campaign of misinformation. The present chapter, "The Dawn of GM Humans," describes the advent of some of the anticipated technologies, their promise, their limitations, and their challenges, both technical and social. How the political dynamics surrounding some of these technologies underpinned the creation of CIRM is the subject of the two chapters that follow, "California Cloning: The Campaign" and "California Cloning: The Aftermath."

The Gene Bubble

It was not just any news item for the daily broadcast. The June 2000 announcement that a nearly complete sequence of the chemical "letters" of the human genome had been deciphered was a well-orchestrated event, special meaning folded into each element. President Clinton led the international debriefing, speaking from the East Room of the White House. Ambassadors from around the globe were in attendance while British Prime Minister Tony Blair stood by waiting to be piped in via satellite. As Clinton explained:

> More than 1,000 researchers across six nations have revealed nearly all 3 billion letters of our miraculous genetic code . . . . Today, we are learning the language in which God created life. We are gaining ever more awe for the complexity, the beauty, the wonder of God's most divine and sacred gift.

But while the genome was proclaimed a divine gift, its sequencing was a human triumph, the creation of "the most important, most wondrous map ever produced by humankind." Clinton drew on the idea of mapping explicitly, underscoring how they were gathered deliberately in the room where Meriwether Lewis had unfolded his map of the American frontier before a wonderstruck Thomas Jefferson, thankful that he had lived to see it. The ordering of the President's thanks also held significance. He structured for himself a clever flanking. Director of the International Human Genome Project, Francis Collins, was paired with Craig Venter, whose biotech company, Celera, had sped against the international consortium ending the race in a tie. It was a civil and corporate framing of an announcement of a scientific feat with religious overtones, decoding the language of God. This cunning launch for the new millennium smoothed reception from all social sectors, even managing those wary of how the new genetic information might be used: "[A]s we consider how to use [this] new discovery," Clinton cautioned, "we must . . . not retreat from our oldest and most cherished human values" (*New York Times*, 2000).

The quest to sequence the human genome, underway for years, nurtured within some scientific and science-watching communities a convenient amnesia, disassociating eugenic dreams from Holocaust nightmares. Could the recurring impulse to improve the human species biologically really be pernicious if individual choice rather than hateful tyranny was the driving instinct? Science luminary James Watson, for example, answered with a resounding, "no." Watson, head of the Human Genome project from 1990–1992, was a co-discoverer in the early 1950s of the famed double helix structure of the DNA molecule. The mesmerizing elegance of DNA's entwined spirals of nucleotide beads, A, C, G, T – adenine (A), cytosine (C), guanine (G), and thymine (T) – became for the biotech era a ubiquitous symbol. Watson's identification with that icon purchased for him the confidence to spew casual remarks that observers and promoters of biotechnology may have found incautious – or perhaps, daring. At a 1998 UCLA symposium Watson, questioned rhetorically: "if we could make better human beings by knowing how to add genes, why shouldn't we?" (Brave, 2003). "[I]f you really are stupid," the Nobel Laureate opined in 2003, "I would call that a disease" (Bhattacharya, 2003). He clarified that he considered 10% of children to be stupid and would like to see them genetically modified. His hopes for tinkering went even further: "People say it would be terrible if we made all girls pretty. I think it would be great," he declared (Stein, 2007).

The Human Genome Project, conceptualized in the 1980s and underway by 1990, seemed to endow the eugenic inclinations of Watson and likeminded colleagues with plausibility. But the results of that project and that of other research in molecular and developmental biology were proving something quite different from what was expected. They were, in fact, relentlessly dismantling the genetic determinist orthodoxy underpinning techno-eugenic desires. Despite President Clinton's embrace of the notion that the human genome sequenced was an etched map, despite media references to its function as life's "blueprint" or how it constituted the "Book of Life," research findings were persistently revealing the unwarranted extravagance of such characterizations. A clear correlation between a gene and a physical trait or condition could not be established in most instances. Moreover, only a small percentage of DNA could be identified as coding for protein production in the cell. Billions of other DNA letters remained a complete mystery. Francis Crick, Watson's acknowledged co-discoverer of DNA's structure, dubbed this chaotic jumble of genetic bits "junk" (Hall, 2012), seemingly unable to imagine for them any value. Results of the Human Genome Project suggested that as much as 97% of the sequence appeared to have no function whatsoever.

Owing to increasing uncertainties about the definition of the gene, debate still rages around this question. Contrary to once prevailing beliefs about how DNA stores information, for example, most coding stretches of DNA specify several proteins with different functions, and these stretches often are controlled by DNA sequences located far from it and from one another, a phenomenon referred to as "**epigenetics**." But whether or not the "rule book" was decipherable by available means, whether or not most DNA sequences were meaningful or largely gibberish, a growing awareness was proving incontrovertible: effects generated within the organism but beyond the gene, and those shaped by the external environment, were becoming prominent explanatory factors for health and disease.

In fact, research was showing that DNA was far from being the "code of life" traditionally claimed for it. Nongenetic influences on embryonic development that can cause even organisms with identical genomes (e.g., twins) to display variable **phenotypes** (physical manifestations) have become well appreciated in recent years (Gordon et al., 2012; Mukherjee, 2016). The confusion that accompanies almost every popular report on genetics is the notion, long rejected by biologists but difficult to completely dispel, that individual genes map one-to-one to specific traits or diseases. But recent work suggests that the problems of genotype–phenotype mapping (i.e., the relation of DNA sequences to the biological characteristics of organisms) go much deeper, to the concept of the gene itself.

Geneticists commonly use terms such as "epigenetics" (narrowly, functional effects from chemical modifications to genes, or more broadly, any external influence on gene activity), "epistasis" (consequences of interaction between the protein or RNA products of different genes), and "incomplete penetrance" (failure of a gene to have its usual effect), to signal their assumption that, apart from such exceptions, a gene, or ensemble of genes will influence a trait in a reliable fashion. But the well-behaved gene may be the exception. A review article in the journal *Human Genetics* discussed the implications of the "many known examples of 'disease-causing mutations' that fail to cause disease in at least a proportion of the individuals who carry them" (Cooper et al., 2013). The authors noted that in some cases the ability of a "bad" gene to cause disease "appears to require the presence of one or more genetic variants at other loci." An unstated implication of this is that whether a gene is pathogenic or not can depend on other atypical genetic variants, and if the "normal" version is substituted in the context of the "wrong" partner genes problems could ensue.

Another study analyzed genomic data from more than 60,000 subjects, focusing on the frequency of mutations previously associated with heritable diseases involving single genes ("Mendelian disorders"). The expectation was that such harmful gene variants would be exceedingly rare, reduced to low levels by natural selection. What was found instead is that that most of them were present too frequently in the broad human population for them to be generally unfavorable. This implies that while they were indeed "bad" genes in the particular subpopulations (containing both affected and unaffected individuals) in which the gene-disease associations were first identified, they are typically innocuous in the human species as a whole (*Nature*, 2016).

The uncertainties of predictive genetics and rational genetic engineering are reflections of the fact that genomes are not machines or computer programs, but have unique evolutionary histories. This can be seen in the very architectures (what is referred to as the basic "body plans") of the animals. Whether animals have ill-defined layers like marine sponges, two layers (ectoderm on the surface and endoderm enclosing the gut) like jellyfish, or three layers (ectoderm, muscle-, bone- and blood-forming mesoderm, and endoderm) like ourselves and insects, or whether their skeletons are inside or outside the body, were first established around 600 million years ago. The genes originally employed in setting the main features of the body – layers and cavities, segments (like our own vertebral bones), muscles, skeletons, nervous systems – continue to have essentially the same functions they started with (Newman, 2016). But over the last half billion years, the functions of these genes have connected and disconnected, integrated and reintegrated, partnered and repartnered with additional genes.

This phenomenon, termed "developmental system drift," has led to the strange state of affairs in which different kinds of animals (e.g., fish, birds, mammals) use the same genes, along with some different ones, to make the "same" structures (backbones, hearts, limbs) in somewhat different ways. Even stranger, members of the same species can use variable genetic means to accomplish the same or similar ends (regulating blood pressure, building a brain). All organisms, including humans, demonstrate this within-species variability (True and Haag, 2001).

A recent genetic survey of a London community of Pakistani origin serves as an example from our own species. The study showed that genes previously thought to be important to normal life processes were inactivated with no obvious effect. In one woman, a gene that determines at what sites the cutting and pasting ("recombination") of chromosomes occurs after fertilization was absent. When this is made to occur in mice (in so-called "gene knockout" experiments), it is fatal to all embryos, and the mutant mice are thus infertile. In contrast, the woman in question had a healthy son, but chromosomal recombination was found to occur at unusual sites. Genes not normally used for this function compensated for the missing one. This suggested that a creative solution occurred in a single generation, rather than as a result of gradual evolutionary refinement (Narasimhan et al., 2016). As reassuring as it is to know that such on-the-spot "saves" are possible, they are exceedingly rare. The take-home message is that the role of a given gene, for good or ill, depends on its context. This is only compounded by the increasing recognition (contrary to a major assumption of genetic science) that the cells and tissues of a single organism exhibit extensive gene variation (Zimmer, 2018).

While a given gene is classically expected to have a similar function wherever it operates, the most serious difficulty in assigning definite functions to genes is that their protein products may not have fixed identities. For more than half a century molecular biology was dominated by the Nobel

Prize-winning doctrine that the polypeptide chains specified by a gene fold in unique fashions, and that the resulting proteins therefore perform similarly in all contexts (Anfinsen, 1973). But, it turns out, this is not always the case. It is now recognized that most proteins involved in cell–cell communication and gene regulation during embryonic development have structures that depend on other proteins that are present in their vicinity (Niklas et al., 2015). These genetically uncoded variations in protein function have even led scientists to identify a protein-based system of inheritance that does not depend on changes in DNA (Chakrabortee et al., 2016).

Although the full implications of these new findings are beyond the scope of this book, they cast serious if not fatal doubt on the feasibility of genetic engineering. The fact that the genetic circuitry of development can be "rewired" over the course of evolution, that given genes function differently in different subgroups within a class of animals (e.g., mammals), and may even do so between individuals within a single species, and that a given gene product (a protein of defined sequence, for example) can function differently in different cells of the same individual, makes "engineering" an organism (in analogy to engineering a skyscraper or a car) an all-but absurd notion. Misbegotten ideas, however, can motivate harmful efforts.

The New Developmental Biology and Dreams of a Scientific Eugenics

The Human Genome Project was accompanied by the deciphering of the genomes of other organisms – the fruit fly and the mouse, among others – which served as "models" for research on biological features shared with humans. The revision of the understanding of the gene and the undermining of its all-powerful role in programming the traits of an organism were largely due to research on such model systems. Nonetheless, the strides in understanding aspects of embryonic development, notwithstanding the more obscure roles of genes, paradoxically invigorated eugenics enthusiasts who envisioned the possibility of manipulating the human embryo. The earliest motivating force behind research on human embryos was simply curiosity about genetic and developmental processes. As little as 40 years ago, molecular genetics and, particularly developmental biology, the field that brought about cloning, embryonic stem cells, and chimerism, had no obvious medical potential or economic prospects. Moreover, research was pursued without specific intention of furthering human genetic engineering per se. It was well understood, however, that these developments profoundly amplified that possibility. Many involved successes with mammalian manipulations. Each widened speculation as to how they might work with the ultimate mammal, the human. Up until this point, the shared genome of the human species had been a product of a common, coherent, evolutionary history. The human genome had been subject to evolution's random shuffling, not to purposeful manipulation or replication. But all this was changing. In the post-1980 world of patent acquisition and profit seeking, this research set the scientific community abuzz with imagined possibilities for application and profit (Parthasarathy, 2017). It also worked in the service of a developing research program for human genetic modification. Four emerging biotechnologies were particularly relevant: cloning, embryonic stem cells, embryo genetic modification, and chimerism. We will discuss the first three of these in the remainder of this chapter and save the last for an extended treatment in Chapter 6, "The Road to Gattaca."

Cloning

The most dramatic and publicly proclaimed development in embryo manipulation was reported in 1997. In February of that year, Scotland's Ian Wilmut and his Roslin Institute colleagues announced the birth of a cloned mammal, a sheep they named, Dolly. To create Dolly, the nucleus of a fresh ovum (egg) taken from one adult sheep was removed and into it was inserted the nucleus of a cell taken from another adult sheep, in this case an adult sheep's mammary cell. Dolly's birth represented a successful attempt following hundreds of repeated failures. Prior implantations and gestations had resulted in seriously deformed newborns. While developing embryos can stay on track, within limits, when perturbed, basic biological principles tell us that a technique bringing together remnants of two damaged cells would not find them cooperating easily to produce a fit member of the species. The technology does not benefit from the error-correcting mechanisms that have evolved over vast periods of time in response to the kinds of cellular damage normally experienced by organisms. Errors in the replication of DNA, for example, a constant of cellular life (unlike the receipt of a foreign nucleus), are repaired by numerous sophisticated enzyme systems. Cloning is therefore a very low-efficiency procedure, with much fetal loss and congenital defects in newborns. While a minority of clones can slip through the shake-out period and survive in a healthy state to old-age, most do not, and the prospects for aged clones depends on the species (Sinclair et al., 2016).

Nevertheless, when a surrogate mother sheep gave birth to Dolly, the technology was declared a success. The lamb was a near identical genetic match to the adult from whom the mammary cell had been taken. Dolly was a clone. A sheep of her breed normally lives 11–12 years. But at six and a half years old, Dolly was euthanized after developing a progressive lung disease and severe arthritis. It is unclear whether the shortened lifespan was directly connected with the sheep's status as a clone, though that is an obvious possibility.

Somatic cell (i.e., derived from the mature body) nuclear transfer or SCNT, as the technology came to be known, despite its record of failures and inefficiency, generated enormous excitement in scientific circles. But among the general public, Dolly generated much unease. She spawned the anxiety that SCNT would be used one day soon to create a cloned human. Research on human embryos had been an issue for some religious groups well before Dolly. In 1994, Catholic political functionary Richard Doerflinger characterized the embryo as a member of the human species, which "should be treated with the same respect accorded any human subject who cannot give consent for himself or herself" (Doerflinger, 1994). But the specter of human clones that Dolly provoked unleashed a surge of secular agitation that triggered immediate political action.

Within a week of Dolly's birth announcement President Clinton issued a moratorium on the use of federal funds for any project involving human cloning and called for voluntary support of the moratorium from researchers using private money (Seelye, 1997). He also asked the National Bioethics Advisory Commission, which he had established two years earlier, to address ethical and legal issues posed by the possibility of human cloning. The Commission's June report concluded that it was morally unacceptable for anyone in the public or private sector, whether in research or clinical setting, to attempt to create a human child using SCNT. Although the Commission emphasized safety, acknowledging that the technology posed significant risks to a potential fetus and/or child, scant consideration was given to the fact that securing the raw material for cloning research would require extracting from women untold numbers of eggs. Its ultimate recommendation to the president was

that the current moratorium on attempts to create children in this manner [SCNT] be continued and that you immediately ask for voluntary compliance in the private sector while federal legislation banning the use of these techniques for creating children is formulated and considered.

(Shapiro, 1997)

But the moratorium did not go unchallenged. In December 1997, physicist Richard Seed announced his intention to create a cloned child. His declaration prompted the Food and Drug Administration to assert its jurisdiction and announce in January 1998 that it would be a violation of federal law to try SCNT to create a child without first obtaining FDA approval (Garvish, 2001). Seed was far from alone in his enthusiasm for full-term cloning (i.e., to create a new human individual). Still, the decision by the Clinton administration to oppose human cloning to full term was never difficult. From the moment the public learned of Dolly's birth, opinion polls consistently documented overwhelming hostility to the idea of human cloning (CGS, 2014). Bio-entrepreneurs, however, felt differently.

Although Wilmut went on record with his hope that no one would attempt to clone a human to full term, the patents awarded to him and his colleagues included human cloning. The reason for seeking patents on human cloning, they explained, was to prevent others from attempting it. Two years after the announcement of Dolly's birth, California's Geron, Inc. held patent rights on embryonic stem cell (ES cell) technology (see Chapter 3). The new company's goal was to generate ES cells from clonal human embryos (Newman, 2003).

As controversy about human cloning brewed, bioentrepreneurs attempted to redefine scientific terminology for public consumption. The goal was to snuff out smoldering contention and popular challenge to professional autonomy. The method was to subdivide the definition of the term "cloning." If the intent in producing a clonal embryo by nuclear transfer was to study it in the hope of finding cures, the recommended term was "therapeutic cloning." If the intent in producing a clonal embryo by nuclear transfer was to bring to term a cloned infant, the recommended term became "reproductive cloning" (Kass et al., 2002; Beverly, 2003). Therapeutic cloning would be acceptable. Reproductive cloning could remain beyond the pale. The new nomenclature did not reflect existing scientific terminology, but appeared to be invented to sow confusion. More specifically, it did not correspond to how scientists in the field described their uses of SCNT. SCNT was the same technology, regardless of whether the researcher's intent was to find cures or create a full-term human clone. Diverting attention away from the technique itself to the intentions of the scientist directing it was a political strategy. The motives of the researcher would function in place of externally imposed regulation.

While many members of the public were troubled by the idea of cloning to create a full-term human genetically identical to the nucleus donor, fewer would be troubled by the idea of cloning to conduct research to find cures. But as we will see in Chapter 6, the changing technological landscape has been accompanied by evolving attitudes on what constitute experiments versus good-faith attempts to produce healthy children. The "three-parent" embryo procedures currently in use to prevent mitochondrial disease and treat infertility, for instance, involve a form of cloning (Newman, 2014).

Bifurcating the terminology in the late '90s eased the way to this "mission creep" in the succeeding decades. It was, as such, an Orwellian tactic, and doubly so: it was a shady parsing of intent and a shrewd invitation to hope. What was being termed "therapeutic cloning" had yet to yield therapies (as it still has not). In the meantime, ES cells have been supplanted to a great extent in the same therapeutic applications by "induced **pluripotent stem cells**" (**iPSC**), which are derived from differentiated tissues of the mature body, even of patients themselves if desired, and do not involve the use of embryos. Methods for producing these cells had not yet been devised when ES cells produced by SCNT were being touted as the only feasible route to tissue repair therapies. They have since garnered a Nobel Prize for their inventor.

As we show later in this book, the acceptance of human cloning as therapeutic strategy has opened the way for its use for reproductive purposes (though not without obfuscation in the form of terminology such as "mitochondrial transfer"). Notwithstanding therapies that eventually derived from ES cells (none of which to date employ SCNT (Lee et al., 2016)), the term "therapeutic cloning" was false advertising. The more accurate term "research cloning" was eventually adopted when the hyperbole was seen to be too much even by bioethicists generally sympathetic to the technology (Pellegrino et al., 2002). But regardless of the specific adjective used, the re-naming project was largely politically successful. It conjured for the public the illusion of a forked road, the distance widening between separate routes of research destined never to meet. It functioned ably to obscure how research undertaken in the hope of finding cures was, in fact, the very same research that enables cloning for reproduction. While not quelling the concerns of some with religious sensibilities about the sanctity of the embryo or of secular critics wary of slippery slopes, it did allay the immediate fears of those discomfited by the prospect of creating viable human clonal embryos for implantation.

There were early opinion makers, including Nathan Myhrvold, the chief technology officer at Microsoft, who enthusiastically embraced the prospect of full-term human cloning (Myhrvold, 1997). Iowa Senator Tom Harkin forecasted favorably in 1997 that, "human cloning will take place, and it will take place in my lifetime." He added, "and I don't fear it at all – I welcome it. I think it's right and proper that we continue this kind of inquiry."

Following the early days after Dolly's birth, enthusiasm on the part of human cloning advocates to create full-term human clones either dampened or went underground. This may be, in part, because of a wider understanding that although cloning results in organisms that are virtually identical genetically, they are not identical in every important respect. One's genes do not fully determine one's traits such as personality, tastes, physical appearance, and susceptibility to disease. More importantly, despite the long-term health of some surviving clones (Sinclair et al., 2016), scientific studies appearing after Dolly's birth continued to confirm what had been suspected beforehand. The procedure produces individuals that are sick from the outset. In all species of animals in which cloning has been attempted, many cloned individuals experience serious anomalies such as grossly abnormal lungs, enlarged hearts, premature aging, as well as high rates of unexplained postnatal deaths.

In 2002 and 2003, Clonaid, a biotech company associated with the Raelian religious cult, claimed to have produced human clones. Although the claims were greeted with skepticism by the press (CNN, 2003), media did not scoff when in November 2007, scientists at Oregon Health and Science University announced that they had created the world's first cloned embryos of monkeys, a species biologically close to humans (Jha, 2007). Months later, in January 2008, scientists at Stemagen Corporation in La Jolla, California announced they had created the first "proven" cloned human embryos (ABC Science, 2008). A decade later animal cloning – cattle, polo ponies (CBS, 2018), companion animals – had become routine. When the entertainer Barbra Streisand cloned her Coton de Tulear, it barely raised an eyebrow (Streisand, 2018).

Although human cloning is against the law in many nations, the fraught politics of human reproduction in the U.S. has prevented any consensus on the matter, and there is thus no federal law that prohibits full-term cloning of humans. Should confirmed reports of relatively healthy human clones come to light, it is not unreasonable to expect that existing opposition to human cloning would decrease. More immediately consequential is the increased demand for clonal human embryos that surged in the wake of the discovery of stem cells. A softening of the objections to creating full-term human clones ticked up in the second decade of the twenty-first century. Technical refinement and social normalization of animal cloning merged with approval (officially in the U.K., with "off-label" uses elsewhere) of **oocyte** nuclear transfer to produce full-term humans. Calls for germline manipulation to improve reproductive outcomes also increased. We discuss all these developments in Chapter 6.

Embryonic Stem Cells and Embryo Germ Cells

In 1981 researchers at the University of Cambridge and the University of California, San Francisco reported a significant result about cells isolated from the **inner cell mass** (**ICM**) of an early mouse embryo. The ICM, a cluster of 100 or so undifferentiated cells, could be caused, by cultivation *in vitro* (outside the body) to become capable of unlimited cell divisions. They became "immortalized." These were **embryonic stem** (**ES**) cells. Embryonic stem cells introduced at an early stage can contribute to all tissues and organs of a developing embryo. When appropriately induced in a culture dish they can differentiate into cartilage, neuron-like cells, the endodermal lining of the gut, and potentially any cell type of the mature body. The announcement, though exciting, was a report on basic research that was far removed from application. Reflecting this, the concluding paragraph of one of the 1981 reports struck an appropriately modest, purely scientific chord: "The availability of such cell lines should make possible new approaches to the study of early mammalian development." But in 1998, when a University of Wisconsin researcher described *human* ES cells, that report's corresponding summary bore the unmistakable commercial stamp of the post-1980 *Chakrabarty*–Bayh-Dole financial revolution: "These cell lines should be useful in human developmental biology, drug discovery, and transplantation medicine." It did not seem to matter that during the 17 years between 1981 and 1998 there had been little talk of any therapeutic potential of embryonic stem cells. In fact, in mouse studies there had been no progress in either curing or easing the symptoms of diseases or disabling conditions.

Significantly, when popular news sources picked up the story in 1998, there was no mention of any of the serious barriers precluding rushing ES cells out of the lab and into human trials. Other research results, for example, were suggesting that when transplanted into a genetically non-identical human host they would most likely provoke an immune reaction that risked killing the patient. Instead, CNN's web headline offered an unproblematic herald: "Researchers isolate human stem cells in the lab: Breakthrough could lead to treatments for paralysis, diabetes" (Newman, 2003). By the time of California's 2004 vote to create and fund CIRM, which prioritized the study of ES cells (see Chapter 3), bioentrepreneurs were hawking the potential of ES cells to cure a host of diseases and disabilities even though there were few experimental studies in the 7 years since their description pointing to their clinical efficacy. In fact, although ES cells could sometimes repair damaged tissues in mice, they also routinely caused malignant tumors. If human ES cells were grafted into human patients, would they behave as mouse ES cells do when they are grafted into mice – would they produce malignant tumors? This remains unclear.

There is another kind of stem cell that does not form tumors when injected into mice, the **embryonic germ stem (EGS)** cell. **Germ cells** produce the developing individual's germline, his or her sperm or eggs. When generated in culture from primordial germ (PG) cells (progenitors of eggs or sperm) isolated from five-to-nine week old late-stage embryos or fetuses EGS cells appear to have a therapeutic potential comparable to that of ES cells in that like the latter they are capable of generating the full spectrum of tissue and cell types. Here too, however, it is not clear how they would behave in human patients. It is likely that since the cells are of a different genotype from the human host they would provoke an immune reaction. The graft could be destroyed or worse, the patient could die (Newman, 2003).

The immune rejection problems that pertained to both ES cells (i.e., those obtained from early-stage embryos) and EGS cells, led to proposals to fund the combining of cloning and stem cell technologies to produce ES cells from clonal embryos (the so-called "therapeutic cloning" described above). Embryos produced by nuclear transfer could be "isogenic" (i.e., having the same genotype) to the patient from whom they were cloned – bringing closer the goal of "therapeutic" cloning. Clonal embryos could also be used to produce isogenic EGS cells. This would depend on the success of research, already underway, on extending the viability of human embryos *in vitro* (i.e., outside a living host, in a dish or test tube) so that such cells could be harvested from two-month old fetuses (Newman, 2003).

Currently, biotech representatives and patient advocacy groups are not promoting the harvesting of EGS cells from fetuses. This could change, however, as research in the area continues to develop. After Dolly was cloned, for example, British embryologist and Bath University professor Jonathan Slack made a proposal that revealed the conceptual framing characteristic of the biotech world that would enable producing "quasi-humans" for such purposes. Slack had recently created headless frogs by manipulating certain embryonically active genes. Recognizing that early work conducted on frogs had led to the birth of the world's first cloned mammal, Dolly, he speculated that research producing headless frog embryos would have implications for mammals as well. Why not produce headless full-term human clones for organ harvesting? "Instead of growing an intact embryo," he suggested, "you could genetically reprogramme the embryo to suppress growth in all the parts of the body except the bits you want, plus a heart and blood circulation." A University College London professor, Lewis Wolpert, seconded the proposal. Wolpert, a senior figure in British biology and prominent public spokesperson on scientific issues, offered the supercilious conclusion that "there are no ethical issues" raised by the procedure (Connor and Cadbury, 1997). These pronouncements were met at the time by derision from many scientists and other commentators. But so was the prospect of full-term, part-human chimeras constructed from cells human and animal embryos when presented via a patent application that same year (Zwerdling, 1998) (see Chapter 6). The fact that construction of such chimeras, which were banned by the NIH as recently as 2015 (NIH, 2015), were approved for funding by the NIH as of 2016 (Kaiser, 2016), indicates the extent of slippage that can occur on attitudes to such questions in a juggernaut-friendly scientific and journalistic environment.

Embryo Gene Modification

Researchers had developed methods to determine the sequence of subunits in DNA molecules by 1977 and, eventually, they identified sequence variations associated with some genetic diseases, e.g., cystic fibrosis and Duchenne muscular dystrophy. For practitioners in the field of reproductive biology, potential applications of sequencing technology for *in vitro fertilization* (IVF, Chapter 1) were obvious. Through a method known as **pre-implantation genetic diagnosis (PGD)**, embryos possessing sequences associated with certain diseases could be identified, thereby giving women the option of not having them implanted.

Development of DNA technology during the following decade, however, brought even more radical possibilities – they crossed the line from eliminating some diseases to the possibility of enhancing offspring. By 1982, researchers succeeded in introducing "foreign" genes into early stage mouse embryos. For the first time in history, such genetically modified or "transgenic" mammals would transmit deliberately introduced genetic information to their offspring. Known as "germline genetic modification," the feat held implications for other species. Would it not be possible one day for humans to birth children possessing introduced gene variants not carried by either parent? And if so, why not all manner of improvements and innovations, as came to be advocated by those now calling themselves "transhumanists" (Bostrom, 2005).

Popular enthusiasts either did not know or did not care that genetically modifying embryos carried health hazards. This is true whether or not modifications are germline (transmitted to future generations) or not. A major germline hazard, for example, is the high rate of tumors that eventually developed in the adult offspring of certain strains of genetically engineered mice (Stewart et al., 1984; Amundadottir et al., 1995). But anomalies appear even in the first generation. Some genetically engineered mice have unexpectedly exhibited limb, brain and craniofacial malformations as well as displacement of the heart (Singh et al., 1991; Griffith et al., 1999).

Developmental intervention techniques can potentially result in anomalies and malfunctions that have yet to be identified in natural populations. The bodies and mentalities of people produced this way would potentially be very different from anyone resulting from sexual reproduction or from "standard" IVF (i.e., IVF not involving embryos that have been genetically engineered). Some parents, desperate for a cure for an existing sick child, have sought to produce a second child, a "savior sibling" from whom they can yield potentially lifesaving bone marrow or umbilical cord stem cells. The inefficient process involves the creation of dozens of embryos to be discarded in search of a good "match" of tissue type, before implantation and birth takes place and the eventual grafted tissue is accepted by the patient. Even then, there is no guarantee of success. Cloning the sick child may increase chances for success because then all embryos created would constitute a perfect match. The first child would not reject the tissue grafted from the second child. In 2002 a mouse study that combined the techniques of embryo gene alteration, ES cells, and cloning, succeeded in doing just that (Rideout et al., 2002), opening the door to eventually creating the so-called savior siblings mentioned above, or potentially down the line, when techniques to ensure the biological product was sufficiently outside the realm of moral concern (as in the Slack-Wolpert proposal), part- or non-reproducing humans for instrumental purposes proposed by the bioengineer Drew Endy (see Specter, 2009 and Chapter 6).

The most widely discussed of recent genetic technologies is the gene "editing" technique, CRISPR/Cas (CRISPR short for "clustered regularly interspaced short palindromic repeats" and Cas, for Cas9, one of several enzymes that can alter DNA in conjunction with specially designed CRISPRs (see Chapter 6). CRISPR (used here as a shorthand) technology makes it possible to snip a DNA sequence at precise points and replace it with another segment. Because it is fast, efficient, relatively accurate, and inexpensive it is touted as the game changer that makes human genetic engineering a robust reality. In March 2015, some of its developers called for a temporary, voluntary moratorium (not a ban) on its use in modifying the human germline in embryos to give the scientific community a chance to acclimate to all its social and ethical ramifications (Stein, 2015). But as news of CRISPR reaches wider circles of public awareness, some of its boosters have worked to neutralize anxiety. An effective, well-tread, if disingenuous, tactic to spin controversy and normalize the technology under consideration is employed: 1) resuscitate genetic determinist thinking and use it to convey a sense of control over genes and predictability about genetic manipulation outcomes, 2) ignore the very basic nature of the state of research and vault to a hoped-for potential of finding cures for an extensive panoply of diseases, infusing expectation with an excitement that hints at imminent remedies, 3) minimize the ways in which the technology is actually different from anything else that society supports or allows, and 4) ignore or downplay the human experimentation involved and the creation of human experimental artifacts that would be necessary to make human genetic engineering at all plausible. In line with this, a "three-parent" embryo construction technique involving piecing together eggs from two different women has been deceptively sold to the British public as "mitochondrial transfer" (see Chapter 6).

George Church is a major promoter of human genetic engineering. He is a co-founder of at least 11 biotech companies, the author of *Regenesis: How Synthetic Biology Will Reinvent Nature and Ourselves* (2012), and a professor of genetics at Harvard Medical School. He holds or has under review more than 70 patents on genetic and related technologies. Church's lab has worked on optimizing CRISPR techniques. In an interview for *New Scientist*, Church waxed enthusiastically about CRISPR's curative potential: "It could enable gene therapies that would allow physicians to fix genetic diseases, including some types of blindness … It could also mean new approaches to treating cancers and viral infections, including HIV" (Griggs and Church, 2015). Implicit in such a forecast of imminent cures is a version of genetic determinism that persists even in the face of that paradigm's revealed oversimplification.

Genetic determinism has been a ubiquitous, false paradigm prevalent since the onset of biotechnology and biotech patenting. Today, though scientists no longer subscribe to the classical version of this notion that theorized straightforward correlations between biological traits and specific genes, and "epigenetics" has become a new watchword, the faith that the complexities of cellular interactions beyond the gene will succumb to human control persists. This by no means implies that basic research involving genes or epigenetics is in decline – it is in fact the reigning approach in the life sciences. But with increasing frequency the refrain in cell and developmental biological articles is, "It was previously believed that gene X just did this, but our new experiments show that it also does that, and that."

Any genetic (or epigenetic) determinist framing of medical applications is the gratuitous pitch of salesmanship. But the blurred boundary between basic research and potentially profitable application makes the determinist thought pattern irresistible to biotechnology promoters. It is uncannily effective in transforming novice listeners into an audience of enthralled true believers.

When asked how he felt about concerns that CRISPR technology constituted a leap closer to creating designer children, Church avoided referring to the perennial over-optimism of genetic determinist projections of the future. Instead, he categorized the germline genetic engineering of offspring as being little different from anything else that parents do for their children: "Do we value long life, intelligence, athleticism, beauty? We are already making our children more educated than their ancestors . . . People pay for products that improve beauty and athleticism" (Griggs and Church, 2015). Church also attempted to normalize CRISPR for human manipulation by comparing its gene modifying potential to other forms of genetic "tinkering" that humans engage in daily:

> We tinker with the gene pool every time we fly at high altitude, which increases random mutations in developing sperm and egg cells. These things are allowed. If you don't like the concept of tinkering with the gene pool, then ban it across the board. Don't be exceptional about CRISPR.
>
> (Griggs and Church, 2015)

It is ironic that his many patent applications have required persuading examiners of exactly the opposite.

Left completely unaddressed by Church are the inevitable unanticipated effects of creating genetically modified children. A 1999 study on mice that were genetically modified to produce a change in their behavioral profile reported that the mice did, as designed, demonstrate superior ability on learning and memory tests (Tang et al., 1999). Popular media referred to the "Doogie" mouse, a reference to a fictional child prodigy on the then popular

TV show, "Doogie Howser, MD." What was not widely reported, however, was that when exposed to chronic stimuli, the mice also exhibited intensified sensation of pain. Few genes have only one effect, and the consequences in a developed person of a gene replacement or intended enhancement introduced when he or she was an embryo cannot be predicted.

Human Modification – for Better or Worse

With what sophistication the general public understood (or understands) the science or social significance of these technologies and their products – cloning, ES cells, and embryo gene modification – remains poorly studied, but in our experience as university teachers in both science and social science it is not high. Certainly, there was reason for optimism when these technologies debuted. But there were grounds for concern as well. Genetically manipulating an early embryo, for example, is unlike modifying a fully formed individual. Early stage manipulation alters an embryo's trajectory and changes it into something intrinsically different from what it would have become. Introduced gene products can have many effects on the tissues and organs taking form during the developmental period, including affecting the wiring of the nervous system and the brain. The bodies and minds of individuals created by developmental intervention could be very different from those generated by evolutionary processes. Even the species-character of a human embryo, once altered, cannot be guaranteed. Although some manipulations may result in "improved humans," either by intent or by accident others will result in entities that are quasi-human or human only by biological affinity. The technologies that brought us the capacity to manipulate the human embryo at its early stages, including its DNA, are placing our common humanity up for grabs. Clearly a world in which genetic manipulations routinely create experimental results ranging from "improved" to less than human "bioproducts" is not the same as a world in which designer children are safely and routinely created.

Promoted by journalists and scientists of a reductionist bent, popular misconceptions about genetic possibilities remain erroneously wedded to the idea that genes are exclusively determinative of biological characteristics. Scientists in the relevant fields are aware that controversial issues must be managed publicly to secure funding and to ensure that research agendas proceed unimpeded. Public relations tactics of bioentrepreneurs concerning genetic technologies include misdirecting public attention away from unavoidable human experimentation by emphasizing a (hoped for) predictability and therapeutic benefit. Unlike the scientists of the postwar generation, especially those influenced by the responsible science movement, few developers share concerns directly with the public. The rare, industry-led calls for moratoria function in self-protective fashion, ultimately normalizing controversial technologies, shepherding them into acceptance. More typically, technologies presenting moral quandaries are channeled through elite bioethics panels, boards, and commissions where members downplay the substantive issues and focus on procedural issues, giving the public a false sense that their interests are being addressed (see Chapter 6). When scientists, or their commercial and bioethicist avatars, *have* addressed the public directly about these technologies, like George Church, their posture has been confident and promotional, either hyping the promise of imminent cures or projecting future scenarios where designer children can be concocted from a catalogue of traits, as quotes opening this chapter exemplify.

In this chapter we have reviewed some of the emerging biotechnologies that hold great consequence for the future of human biology and society, and it provokes crucial questions. Should bioentrepreneurs, in whose immediate interest it is to see these biotechnologies advanced, be left to characterize the safety, ethical implications, and possible social outcomes and fast track their implementation? Or should such biotechnologies be subjected to comprehensive democratic consideration and oversight? These questions are of particular concern in part because technical barriers exist that make public understanding and involvement challenging. There is almost no public understanding, for example, of the difference between somatic (mature body cell) genetic modification and embryo modification. The first (such as the genetic experimentation involving the patient Jesse Gelsinger, and the recent anti-blindness efforts described in Chapter 1), though shamelessly hyped and in some cases irresponsibly implemented, may in fact eventually yield treatments for otherwise incurable conditions. The second (discussed in this chapter), directed to designing people who do not yet exist, would commodify human reproduction, most likely create more health problems than it solves, and alter the engineered individual's germ (egg or sperm) cells, potentially passing experimental errors to future generations. Currently, it is difficult for citizens to get a firm grasp of what is at stake in part because bioentrepreneurial commercial, political, and public relations practices work to prevent clear-sighted understanding. In the next chapter we present a case study, passage of the California Stem Cell Research and Cures Act, to explore some of these practices.

Sources Consulted for Chapter 2

ABC Science, "First 'Proven' Human Cloned Embryo," January 18, 2008: www.abc.net.au/science/articles/2008/01/18/2141478.htm

Amundadottir, L.T., M.D. Johnson, G. Merlino, G.H. Smith, and R.B. Dickson, "Synergistic Interaction of Transforming Growth Factor Alpha and C-Myc in Mouse Mammary and Salivary Gland Tumorigenesis," *Cell Growth Differ* 6(6) (1995): 737–748.

Anfinsen, C.B., "Principles that Govern the Folding of Protein Chains," *Science* 181 (1973): 223–230.

Beeson, Diane, and M.L. Tina Stevens, "Big Biotech and Abortion Politics: The Progressive Campaign Against California's 2004 Stem Cell Initiative," 2004, PDF available upon request from authors or publisher.

Beverly, Blair, "Stemming Controversy: Developing Stanford's Institute for Cancer/Stem Cell Biology and Medicine," *The Stanford Scientific* (Spring 2003): 28–30.

Bhattacharya, Shaoni, "Stupidity Should Be Cured, Says DNA Discoverer," *The New Scientist*, February 28, 2003: www.newscientist.com/article/dn3451-stupidity-should-be-cured-says-dna-discoverer/

Bostrom, N., "A History of Transhumanist Thought," *Journal of Evolution and Technology* 14(1) (April 2005): 1–25.

Brave, Ralph, "James Watson Wants to Build a Better Human," *Alternet*, May 28, 2003: www.alternet.org/story/16026/james_watson_wants_to_build_a_better_human

Center for Genetics and Society, "CGS Summary of Public Opinion Polls," February 4, 2014: www.geneticsandsociety.org/article.php?id=401

Chakrabortee, S., J.S. Byers, S. Jones, D.M. Garcia, B. Bhullar, A. Chang, R. She, L. Lee, B. Fremin, S. Lindquist, and D.F. Jarosz, "Intrinsically Disordered Proteins Drive Emergence and Inheritance of Biological Traits," *Cell* 167 (2016): 369–381.

Church, George, *Regenesis: How Synthetic Biology Will Reinvent Nature and Ourselves*, New York: Basic Books, 2012.

CNN.com, "Clonaid Says It's Cloned First Boy," January 23, 2003: www.cnn.com/2003/WORLD/americas/01/23/clonaid.claim/

Connor, Steve, and Deborah Cadbury, "Headless Frog Opens Way for Human Organ Factory," *London Sunday Times*, October 19, 1997: www.organicconsumers.org/old_articles/Patent/headless.html

Cooper, D.N., M. Krawczak, C. Polychronakos, C. Tyler-Smith, and H. Kehrer-Sawatzki, "Where Genotype Is Not Predictive of Phenotype: Towards an Understanding of the Molecular Basis of Reduced Penetrance in Human Inherited Disease," *Hum Genet* 132 (2013): 1077–1130.

Doerflinger, Richard, *Public Comment: NIH Human Embryo Research Panel*, February 2, 1994: www.usccb.org/prolife/issues/bioethic/embryo/rd20294.htm

Garvish, John, "The Clone Wars: The Growing Debate over Federal Cloning Legislation," *Duke Law & Technology Review* 0022 (June 20, 2001): 4. www.law.duke.edu/journals/dltr/articles/2001dltr0022.html.

Gordon, L., J.E. Joo, J.E. Powell, M. Ollikainen, B. Novakovic, X. Li, R. Andronikos, M.N. Cruickshank, K.N. Conneely, A.K. Smith, R.S. Alisch, R. Morley, P.M. Visscher, J.M. Craig, and R. Saffery, "Neonatal DNA Methylation Profile in Human Twins Is Specified by a Complex Interplay between Intrauterine Environmental and Genetic Factors, Subject to Tissue-Specific Influence," *Genome Res* 22 (2012): 1395–1406.

Griffith, A.J., W. Ji, M.E. Prince, R.A. Altschuler, and M.H. Meisler, "Optic, Olfactory, and Vestibular Dysmorphogenesis in the Homozygous Mouse Insertional Mutant Tg9257," *J Craniofac Genet Dev Biol* 19(3) (1999): 157–163.

Griggs, Jessica and George Church, "Our Superhuman Future Is Just a Few Edits Away," *The New Scientist* (September 25, 2015): 28–30.

Hall, Stephen S., "Hidden Treasures in Junk DNA," *Scientific American*, September 18, 2012: www.scientificamerican.com/article/hidden-treasures-in-junk-dna/

Jha, Alok, "Scientists Create World's First Monkey Embryos," *The Guardian*, November 14, 2007: www.theguardian.com/science/2007/nov/14/clone.stem.cells

Kaiser, Jocelyn, "NIH Moves to Lift Moratorium on Animal–Human Chimera Research," *Science*, August 4, 2016: www.sciencemag.org/news/2016/08/nih-moves-lift-moratorium-animal-human-chimera-research

Kass, Leon, et al., "Human Cloning and Human Dignity: The Report of the President's Council on Bioethics," *New York: Public Affairs* (2002): 41–63.

Lee, J.E., Y.G. Chung, J.H. Eum, Y. Lee, and D.R. Lee, "An Efficient SCNT Technology for the Establishment of Personalized and Public Human Pluripotent Stem Cell Banks," *BMB Rep* 49 (2016): 197–198.

Mukherjee, Siddhartha, "Same but Different: How Epigenetics Can Blur the Line between Nature and Nurture," May 2, 2016: www.newyorker.com/magazine/2016/05/02/breakthroughs-in-epigenetics

Myhrvold, N., "Human Clones: Why Not?" *Slate*, 1997: www.slate.com/articles/briefing/critical_mass/1997/03/human_clones_why_not.html

Narasimhan, V.M., K.A. Hunt, D. Mason, C.L. Baker, K. J. Karczewski, M.R. Barnes, A.H. Barnett, C. Bates, S. Bellary, N.A. Bockett, K. Giorda, C.J. Griffiths, H. Hemingway, Z. Jia, M.A. Kelly, H.A. Khawaja, M. Lek, S. McCarthy, R. McEachan, A. O'Donnell-Luria, K. Paigen, C.A. Parisinos, E. Sheridan, L. Southgate, L. Tee, M. Thomas, Y. Xue, M. Schnall-Levin, P.M. Petkov, C. Tyler-Smith, E.R. Maher, R.C. Trembath, D.G. MacArthur, J. Wright, R. Durbin, and D.A. van Heel, "Health and Population Effects of Rare Gene Knockouts in Adult Humans with Related Parents," *Science* 352 (6284) (2016): 474–477.

Nature, Editorial, "Genetic Reckoning: Researchers Need to Reassess Many Accepted Links between Mutations and Disease," 538(140) (October 13, 2016).

Newman, Stuart A., "Averting the Clone Age: Prospects and Perils of Human Developmental Manipulation," *Journal of Contemporary Health Law, and Policy* 19 (2003): 431–463.

Newman, Stuart A., "Origination, Variation, and Conservation of Animal Body Plan Development," *Reviews in Cell Biology and Molecular Medicine* 2 (2016): 130–162.

New York Times, "Reading the Book of Life," www.nytimes.com/2000/06/27/science/reading-the-book-of-life-white-house-remarks-on-decoding-of-genome.html

NIH, "Research Involving Introduction of Human Pluripotent Cells into Non-Human Vertebrate Animal Pre-Gastrulation Embryos," September 23, 2015: http://grants.nih.gov/grants/guide/notice-files/NOT-OD-15-158.html

Niklas, K.J., S.E. Bondos, A.K. Dunker, and S.A. Newman, "Rethinking Gene Regulatory Networks in Light of Alternative Splicing, Intrinsically Disordered Protein Domains, and Post-Translational Modifications," *Front Cell Dev Biol* 3(8) (2015). https://core.ac.uk/download/pdf/82839484.pdf

Parthasarathy, Shobita, *Patent Politics: Life Forms, Markets and the Public Interest in the United States and Europe*, Chicago, IL: University of Chicago Press, 2017.

Pellegrino, Edmund, D., John F. Kilner, Kevin T. FitzGerald, Linda K. Bevington, C. Ben Mitchell, and C. Everett Koop, "Therapeutic Cloning" (Correspondence) *N Engl J Med* 347 (November 14, 2002): 1619–1622: www.nejm.org/doi/full/10.1056/NEJM200211143472014#t=article

Rideout, W.M. 3rd, K. Hochedlinger, M. Kyba, G.Q. Daley, and R. Jaenisch, "Correction of a Genetic Defect by Nuclear Transplantation and Combined Cell and Gene Therapy," *Cell* 109(17) (2002): 23.

"Scientist Who Cloned Sheep: Cloning Would Be Inhuman," *CNN*, March 12, 1997: www.cnn.com/HEALTH/9703/12/nfm/cloning/

Seelye, Katharine Q., "Clinton Bans Federal Money for Efforts to Clone Humans," *New York Times*, Mar 5, 1997: www.nytimes.com/books/97/12/28/home/0305clinton-cloning.html?

Shapiro, Harold T. et al., "Cloning Human Beings: Report and Recommendations of the National Bioethics Advisory Commission," Rockville, MD, June 1997: www.georgetown.edu/research/nrcbl/nbac/pubs/cloning1/cloning.pdf

Silver, Lee, *Remaking Eden: How Genetic Engineering and Cloning Will Transform the American Family*, Toronto: Harper Collins, 1997.

Sinclair, K.D., S.A. Corr, C.G. Gutierrez, P.A. Fisher, J.H. Lee, A.J. Rathbone, I. Choi, K.H. Campbell, and D.S. Gardner, "Healthy Ageing of Cloned Sheep," *Nature Communications* 7(12359): www.nature.com/ncomms/2016/160726/ncomms12359/full/ncomms12359.html

Singh, G., D.M. Supp, C. Schreiner, J. McNeish, H.J. Merker, N.G. Copeland, N.A. Jenkins, S.S. Potter, and W. Scott, "Legless Insertional Mutation: Morphological, Molecular, and Genetic Characterization," *Genes Dev* 5 (12A) (1991): 2245–2255.

Specter, M., "A Life of Its Own: Where Will Synthetic Biology Lead Us?," *The New Yorker*, September 28, 2009: 56–65.

Stein, Lisa, "Nobel Scientist Quits in Wake of Scandal," *Scientific American*, October 25, 2007: www.scientificamerican.com/article/nobel-scientist-quits-in/

Stein, Rob, "Scientists Urge Temporary Moratorium On Human Genome Edits," NPR, *All Things Considered*, March 20, 2015: www.npr.org/sections/health-shots/2015/03/20/394311141/scientists-urge-temporary-moratorium-on-human-genome-edits

Stewart, T.A., P.K. Pattengale, and P. Leder, "Spontaneous Mammary Adenocarcinomas in Transgenic Mice That Carry and Express MTV/Myc Fusion Genes," *Cell* 38(3) (1984): 627–637.

Stock, Gregory, *Redesigning Humans: Our Inevitable Genetic Future*, Boston, MA: Houghton Mifflin, 2002.

Streisand, Barbara, "Barbra Streisand Explains: Why I Cloned My Dog," *The New York Times*, March 2, 2018: www.nytimes.com/2018/03/02/style/barbra-streisand-cloned-her-dog.html

Tang, Y.P., E. Shimizu, G.R. Dube, C. Rampon, G.A. Kerchner, M. Zhuo, G. Liu, and J.Z. Tsien, "Genetic Enhancement of Learning and Memory in Mice," *Nature* 401 (1999): 63–69.

True, J.R., and E.S. Haag, "Developmental System Drift and Flexibility in Evolutionary Trajectories," *Evol Dev* 3 (2001): 109–119.

www.nytimes.com/2015/12/04/science/crispr-cas9-human-genome-editing-moratorium.html?_r=0

Weiner, Charles, "Social Responsibility in Genetic Engineering: Historical Perspectives," Anders Nordgren, ed. *Gene Therapy and Ethics: Studies in Bioethics and Research Ethics* 4, Uppsala: Acta Universitatis Upsaliensis, 1999, 51–64.

Zimmer, Carl, "Every Cell in Your Body Has the Same DNA. Except It Doesn't." *New York Times*, May 21, 2018: www.nytimes.com/2018/05/21/science/mosaicism-dna-genome-cancer.html

Zwerdling, D., "Humanimals. All Things Considered," National Public Radio, USA, April 5, 1998: www.npr.org/news/healthsci/indexarchives/1998/apr/980405.02.html

3
California Cloning
The Campaign[1]

[A] mix of sexual and clonal reproduction makes good sense for genetic design. Leave sexual reproduction for experimental purposes; when a suitable type is ascertained take care to maintain it by clonal propagation . . .. From a strictly biological standpoint, tempered clonality would allow the best of both worlds: we would at least enjoy being able to observe the experiment of discovering whether a second Einstein would outdo the first one. How to temper the process and the accompanying social frictions is another problem.

(Joshua Lederberg Nobel Laureate, Stanford molecular biologist from, "Experimental Genetics and Human Evolution," 1966, pp. 527–528)

[for Lederberg] . . . serious talk about cloning is essentially crying wolf when a tiger is already inside the walls. This position, however, fails to allow for what I believe will be a frenetic rush to do experimental manipulation with human eggs once they have become a readily available commodity.

(James Watson, Nobel Laureate from, "Moving Toward the Clonal Man," *The Atlantic*, 1971, p. 52)

We can clone all kinds of mammals, so it's very likely that we could clone a human. Why shouldn't we be able to do so?

(George Church, Professor of Genetics, Harvard Medical School, bioentrepreneur, quoted by Smith, 2013, www.telegraph.co.uk/news/science/ 9814620/I-can-create-Neanderthal-baby-I-just-need-willing-woman.html)

It was fall 2004 and California was gearing up for its November general election. Sporting white lab jackets and their university affiliations, bioentrepreneurs took to the state's commercial airwaves, spoke at rallies, and offered quotes for thousands of glossy brochures. Their goal was to secure vast public resources for a speculative and controversial line of research left largely unfunded by the federal government. It was part of an audacious, unprecedented strategy to finance and develop policy on complex scientific issues by going directly to voters. Backed by a $34 million war chest (the largest sum ever spent on a California ballot proposition) they were banking on the hopes of people seeking relief from a wide range of diseases and disabling conditions. Surely, voters would pass a state initiative that promised imminent cures, Proposition 71. "Passage of The California Stem Cell Research and Cures Initiative [Prop 71] will energize vitally needed research . . . for the use of stem cells to cure millions of children and adults," promised Stanford University's Paul Berg, a 1980 Nobel Prize winner in Chemistry (Yes on 71 Archive).

The sheer volume of promotional material and its air of urgency evoked the sense that breakthroughs were close at hand. The "Yes on 71" website featured a "countdown to cures" graphic (eventually removed) and claimed that, "128 million Americans suffer from diseases and injuries that could be treated or cured with stem cell therapies." Perhaps the most guileful endorsement assured that, "Voting Yes on 71 could save the life of someone you love." Even the campaign's contact information, info@curesforcalifornia.com or toll free call line at 1-800-931-CURE, telegraphed the irresistible hypnotic mantra: cures, cures, cures. Moreover, promoters assured, these cures would save Californians an immense sum otherwise spent on healthcare: "If Prop. 71 leads to cures that reduce our health care costs by only 1%, it will pay for itself – and it could cut health care costs by tens of billions of dollars in future decades" (UCLA Digital Library, Prop 71). Reportage aired the proposition backers' optimistic assumptions that research would result in medical advances for at least six diseases within 15 years (Cohen, 2005).

Bullish assurances offered at the 2004 National Democratic Convention buoyed California voters. There, Ron Reagan, son of the former president then suffering from Alzheimer's disease, told the teeming primetime crowd that stem cells would result in "your own personal biological repair kit standing by at the hospital." His extraordinary promise capped a parade of convention orators wedding their political futures to science prophecy (Witherspoon Council, 2012). Saturated with promises of imminent healing and ultimate savings, Californians voted 59% to 41% to pass into law the California Stem Cell Research and Cures Bond Act which created the California Institute of Regenerative Medicine (CIRM) to administer the largess (Ballotpedia, 2004).

According to the terms of Prop 71, CIRM would be unlike other state agencies. It would be essentially unanswerable to the legislature or the governor, and exempt from state laws relating to good governance, including open meeting laws. Veteran California journalist David Jensen's well-regarded blog, the *California Stem Cell Report*, strives to maximize transparency of CIRM. Jensen correctly characterized the agency as functioning, "almost invisibly with little attention from any media, mainstream or otherwise" (CSCR, Oct. 25, 2017). More than a dozen years on, CIRM has spent all the $3 billion gifted to it by voters, with a mixed record of accomplishments (discussed in Chapter 4). An additional $3 billion in interest on the bonds raises the massive debt burden for the state to $6 billion. CIRM moved its headquarters to Oakland, its ten years of free rent in one of San Francisco's most coveted commercial regions having run out (Colliver, 2015). In the meantime, its countdown continues. Unless new funding appears, private, public or both, it will be a countdown to termination.

It is still uncertain whether CIRM may find or make widely available any cures that might come from research that it funds. Any such development would of course be welcome news. But CIRM is a new interest group, comprised of the beneficiaries of public largesse. It can be expected to act in its own interests to consolidate its power and its access to funding. Robust vigilance against exaggerated promises is therefore essential. This chapter focuses on three problematic areas: first, the strategies employed by entrepreneurial interests to secure and control funding from the State of California for controversial bioscience research; second, the conflicts of interest buttressing some of these strategies; and third, the buried social and ethical quandaries that ride in tandem with the research. In so doing, the chapter prompts scrutiny of seeking voter based state funding of biotechnology. How possible is it to enforce effective transparency and accountability when motivations pushing basic research toward human application braid so tightly with strands of career building and profit seeking? The chapter assays these issues by relating the struggle of prochoice progressive activists to cast the light of day on highly significant yet skillfully hidden social issues associated with the bioresearch: human egg harvesting and human cloning. The story provokes a final question: can a humane society and a human future be secured if such consequential points of departure remain clouded?

Clouding Issues at the Origin of CIRM

The text of Prop 71 obscured how the initiative stood to prioritize human embryonic stem cell (hES) research, especially a type of human embryonic cloning research, **somatic cell nuclear transfer (SCNT)**. Because it involves the destruction of human embryos, national abortion politics made federal funding of this research largely insupportable. Starting in 2001, the Bush administration would permit funding for research on existing embryonic stem cell lines that had already been derived from embryos, but not the creation of new ones (Murugan, 2009). Prop 71's authors avoided the words "embryo" or "cloning" in their description of the prioritized research and used instead words unfamiliar to the general public that referenced the cells derived from clonal embryos but not the sources of these cells: It is the intent of the people of California in enacting this measure to:

> [m]aximize the use of research funds by giving priority to stem cell research that has the greatest potential for therapies and cures, specifically focused on pluripotent stem cell and progenitor cell research among other vital research opportunities that cannot, or are unlikely to, receive timely or sufficient federal funding, unencumbered by limitations that would impede the research.
>
> (Voter Information Guide, 2004)

Progressive prochoice science watchers, not motivated by religious sensibilities about the moral status of embryos, recognized two issues that were nonetheless socially troubling. The first was that the research required human eggs, thereby appealing to young women to subject themselves to egg harvesting and its understudied health risks. The second was that the research, though not explicitly concerned with the creation of full-term human clones, does advance human cloning research as an adjunct to producing patient-compatible stem cells (discussed later in this section), as well as embryo "disease models." The latter are human embryos produced by SCNT using nuclei extracted from patients with gene-associated diseases (e.g., muscular dystrophy, cystic fibrosis). The objective is to study how the disease takes form at the early stages of development.

The year before the passage of Prop 71, California prohibited full-term human cloning (Beeson and Stevens, 2004), but there was and remains no federal ban. Embryos created in California, even by SCNT (i.e., cloning), may be implanted for purposes of gestating them in any of the dozen or so states or approximately 40 countries that does not specifically prohibit it. California voters would find little to enlighten them on these controversial issues in the text of Prop 71 in their *Official Voter Information Guide*. In her account of Prop 71's passage, UC Berkeley professor Charis Thompson characterized succinctly what voters never learned from the text of the initiative they had passed:

> a constitutional right was established to conduct human embryonic stem cell research in the state of California, and three billion dollars of taxpayer money was pledged to support the research and its real estate requirements, without mentioning women, embryos, eggs, research donors . . . or research cloning.
>
> (Thompson, 2013, p. 87)

The California Stem Cell Research and Cures Initiative as it appeared in the guide was over eight pages of dense, small font text peppered with scientific jargon concealing far more consequential information than it disclosed. Like the campaign itself and the multi-million dollar celebrity-studded media blitz designed to promote it, Prop 71 opened by heralding the potential for cures. "Millions of children and adults suffer devastating diseases that are currently incurable, including cancer," the text began. An impressive list of additional diseases and conditions that Californians could help cure if they supported the Proposition followed: "diabetes, heart disease, Alzheimer's, Parkinson's, spinal cord injuries, blindness, Lou Gehrig's disease, HIV/AIDS, mental health disorders, multiple sclerosis, Huntington's disease, and more than 70 other diseases and injuries." How could these cures be realized? The answer elevated an idealized cell type to panacea status: "The cure and treatment of these diseases can potentially be accomplished through the use of new regenerative medical therapies including a special type of human cells, called stem cells" (Voter Information Guide, 2004).

In fact, there are different types of stem cells, some of which had resulted in demonstrable benefits, others that had not, at least as of 2004. Even so, the text of Prop 71 referred generally to stem cells and made generic claims about their therapeutic attributes. The accompanying legislative analysis also reported generically that stem cell research had "led to the development of treatments of a variety of cancers and blood disorders." But, in reality, this claim could only be sustained for "adult" stem cells. Adult stem cells are found among differentiated cells in organ and tissue. Although nestled there, they are themselves undifferentiated, that is, they have not developed into a specialized cell type. They can renew themselves, they are capable of differentiating into specialized cell types, and at the time Prop 71 was being advanced, had led to some notable treatments. Bone marrow transplants, for example, consist of mixtures of different types of adult stem cells. Funding of adult stem cell research has never been off the table for federal agencies. Thus, although ES cell research was unnamed as such in the text, it was clearly Prop 71's motivation, as indicated by the language suggesting prioritization of federally unfunded pluripotent and progenitor cell research.

The research on human ES cells was inevitably speculative. But even demonstrations of therapeutic potential in mice, where ES cells had already been studied for more than 20 years, were meager (Wobus and Boheler, 2005). In fact, prior to the passage of Proposition 71 venture capitalists were pulling away from the area. A *USA Today* markets and stocks journalist reported in 2001 that "investors see stem-cell companies as nothing more than money-burning firms and are treating them like the plague" (Krantz, 2001). Two decades of work on human ES cells have resulted in only a few hints of treatments or cures. These cells also exhibit "tumorigenicity" – that is, they cause cancer, a problem that can be controlled, but is still a concern (Priest et al., 2015). Prop 71 promoters, however, never revealed to voters this highly significant hazard. The text assured Californians that the only obstacle to cures was a financial one: "life-saving medical breakthroughs can only happen if adequate funding is made available to advance stem cell research, develop therapies, and conduct clinical trials" (Official Voter Info Guide, Text of Proposed Laws, Sec. 2, p. 147). Why not be fully forthcoming about the categories and properties of stem cells?

First, as noted, deriving ES cells necessitated the destruction of embryos, which some religious groups, considering all stages of human life to be morally equivalent, objected to on principle. The federal funding ping-pong ball dynamic on this issue between the Democratic and Republican administrations under Presidents Clinton and Bush underscores its contentiousness. Convinced that ES cell research was a promising path to curing disease, President Clinton favored ES cell research using existing embryos. He was constrained from providing federal funds for it, however, by the Dickey-Wicker Amendment, a rider attached to the 1995 appropriations bill for the Department of Health and Human Services (named for its Republican authors, Arkansas' Jay Dickey and Mississippi's Roger Wicker). But a legal opinion by the Department of Health and Human Services offered a way around this restriction. It declared that already existing stem cells "are not a human embryo." This enabled NIH to begin accepting grant proposals for embryonic stem cell research. Eight months after President George W. Bush took office in 2001, however, he announced that federal funding would be restricted to already existing stem cell lines. For the foreseeable future, no federal funds would be forthcoming to create new embryonic stem cells lines. Indeed, this religiously influenced twist of the funding stream (reversed once the pendulum shifted back to a Democratic administration) was the driving impetus for devising Prop 71. The day after the California election, the Stanford news service referred to passage of Prop 71 triumphantly as "a direct rebuke to the federal government" (Adams, 2004). The following year, a writer for *Smithsonian Magazine* characterized Prop 71 as a "referendum" (although it was actually an initiative, a very different legal instrument) that "sent a blaring signal to Washington." "Prop 71," he opined, "is . . . almost a declaration of secession" (Cohen, 2005).

Second, the embryos used in the prospective research would typically be "clonal," in the "low-tech" sense that Aldous Huxley described in his 1931 novel *Brave New World*. That is, the clusters of up to 16 cells (called "**blastomeres**") that constitute early embryos would be separated from one another in vitro, and the individual cells would then be induced to make genetically identical new embryos (like identical twins); these could be separated in turn to make even more embryos. Acceptance of this method by the research community would ease the way to the next step, generating human embryos for research by a different kind of cloning, SCNT (see Chapter 2), which involved replication of existing genetic prototypes. Both of these cloning techniques are currently in use for the production of cattle and other animals. (In the nonhuman applications, of course, the clones are brought to full term.) The initial rationale for SCNT cloning of humans would be the desirability of studying the known diseases of selected nucleus donors. When (if the biotech juggernaut remains on track) it is eventually added to the IVF industry's toolbox (which already is being openly discussed in China) (RT, 2015), the social and ethical quagmires accompanying the manipulation of embryos, namely, the creation of human clones and the facilitation of "designer" children would be unleashed (see Chapter 2). If the old-fashioned technique of subdividing embryos is the dystopic governmental-control scenario of *Brave New World*, SCNT cloning opens the door to the biohacker visions of the 1970s novel and film, *The Boys From Brazil* and the 2013–2017 Canadian TV drama, *Orphan Black*.

Finally, the raw material required for creating clonal embryos, women's ova (eggs), was not easily acquired and there are inadequately examined health consequences for women supplying them. For many years women's health advocates had been concerned by the social and health effects of understudied reproductive technologies (Arditti et al., 1984) and a steady stream of anecdotal reports of very serious short and long-term health problems in egg providers resulting not only from the surgery required to remove the eggs, but also from their exposure to large amounts of synthetic hormones (Norsigian, 2005). As the demand for women's eggs for research swelled, eggs needed now not only for IVF treatment but also for SCNT, concern intensified (Norsigian and Newman, 2001). Researchers would need thousands of eggs for the proposed studies. Pharmaceutical companies had not been required to collect safety data on the powerful hormones used to stimulate egg production to bring abnormally large numbers of eggs to maturity for their removal (Parisian, 2005). The short and long-term effects of ovarian stimulation were, and remain, concerning. Short-term risks include **Ovarian Hyperstimulation Syndrome (OHSS)** with outcomes ranging from nausea, vomiting, and distention, to (much less commonly, though way too often for complacency and almost certainly under-reported) organ failure and death. The few studies completed on long-term effects of ovarian stimulation reached conflicting conclusions but included cancer.[2]

Some of the drugs used for egg production and extraction, having been approved "off label," meant that the FDA had not required full accountability for how and in which medical contexts they were used. Of particular concern was the drug "Lupron" (leuprolide acetate,) which was being used off label (i.e., it had been approved to treat men with prostate cancer and as a treatment for women with endometriosis but it had not been approved for use in egg harvesting). Some of the women who had developed debilitating symptoms after using the drug created the "Lupron Victim's Network" to share their stories. The FDA had more than 6,000 complaints about the drug on file that had not been investigated. This included 25 deaths. A few months following passage of Prop 71, former Chief Medical Officer of the Food and Drug Administration, Suzanne Parisian, wrote an open letter warning that "those promoting SCNT research may be unknowingly tackling a far more costly and serious health burden by allowing the expanded use of current IVF stimulation drugs for SCNT." Health advocates also were troubled by how financially disadvantaged women were particularly vulnerable to the inducements of thousands of dollars paid for the "donations" of their eggs (Beeson and Lippman, 2006). Lupron also is associated with serious adverse health effects for women who were given the drug as children to delay puberty or to cause them to grow taller (Jewett, 2017). None of these concerns were being addressed.

Enter California

To outflank these concerns and to avoid national wrangling and roadblocks put up by abortion politics, bioentrepreneurs resolved to blaze a political trail to the states. The obvious venue was a state often credited as the birthplace of modern biotechnology and whose electorate was largely prochoice: California. In the wake of the announcement of the birth of the first cloned mammal (the sheep, Dolly), bio-researchers worked to construct a legal environment in California that would court the public and neutralize opposition to the development of cloning technology. An early step was to secure statutory legitimation of terminology that would obscure the cloning aspect of the planned research.

In 1997, California Senate Bill 1344 and Senate Concurrent Resolution 39 banned the purchase or sale of "an ovum, zygote, embryo, or fetus for the purpose of cloning a human being" (California Codes. Health and Safety Code Section 24185c(3)). In the course of so doing, it encoded into law the conceptual distinction between cloning to produce embryos for reproduction and cloning to produce embryos for research. This politically inspired distinction was crucial to the project of evading overwhelming public opposition to human cloning. The goal was to obtain financial support for what was, in fact, cloning research. As described in Chapter 2, there is no difference in the technology employed for "research cloning" and "reproductive cloning." The difference is only in the researcher's intention, that is, what is intended to be done subsequently with the clonal embryo. To organize a public response around this distinction is a departure from the scientific practice of naming things objectively, rather than subjectively.

Research cloning is a *sine qua non* for reproductive cloning. The intense national controversy over politicized scientific terminology was settled in

California in favor of definitions that make it appear to the layperson that these are distinct technologies. This eased the way for the development of cloning for whatever ultimate purpose. It created the illusion of a forked road with separate routes of non-intersecting research. Such an illusion undermines a crucial realization: stopping human reproductive cloning requires more than one state's declaration that it will not permit creation of a full-term human clone. Only national and international agreement could accomplish that. Without it, nothing prevents the research cloning technology being perfected in one state from being put to reproductive uses in another.

California Senate Bill 1344 and Senate Concurrent Resolution 39 also established a five-year moratorium on "the cloning of an entire human being" and called upon the Genetic Disease Branch of the California Department of Health Services to appoint a group of 12 individuals to form the California Advisory Committee on Human Cloning to gather more information and to advise the Governor and the legislature. Although the committee was mandated to represent various perspectives or constituents including those of biotechnology, genetics, medicine, law, ethics, religion, and the public, in fact, it excluded known critics and was heavily weighted toward bioentrepreneurs. Formally appointed in December 1998, the panel issued its report in January 2002.

The Advisory Committee accepted the newly established terminology. And, although not a forgone conclusion, it eventually agreed unanimously that California should prohibit human reproductive cloning and reasonably regulate human non-reproductive cloning (Quintero, 2001). In September 2002, in keeping with the recommendations of the Committee, Governor Gray Davis signed legislation that continued the ban on "reproductive cloning." More importantly, the Committee granted explicit permission to engage in "non-reproductive cloning" (McLean, 2002). This law, California Health and Safety Code Section 125300, defines the policy of the State of California as permitting research involving the derivation and use of human embryonic stem cells, human embryonic germ cells, and human adult stem cells from any source, including SCNT (i.e., cloning). It ran directly counter to federal policy, which severely restricted funding for embryonic stem cell research.

Authorizing funds for the research, however, would prove much more problematic than legalizing it. California's recent energy crisis and the Enron debacle's devastating consequences for the state treasury made funding science research (usually left to the national government) difficult for the state to prioritize (Greenfield, 2004). In September 2003, the Senate Appropriations Committee killed a bill that would have authorized the issuance of bonds for stem cell research, SB 778. It met a similar fate the following legislative session. Its defeat, combined with the lack of any progress at the federal level, set the stage for employing the state initiative process to appeal directly to the electorate in the 2004 general election. This became Proposition 71.

Late in 2002, a quarrel between the new Stanford Cancer/Stem Cell Biology and Medicine Institute and the Bush-appointed President's Council on Bioethics (Novak, 2003) brought forth a cunning strategy for moving ahead: eliminate the term "cloning" altogether. The Institute's director, Irving Weissman, wanted the procedure called "therapeutic cloning" to be referred to instead as "nuclear transplant to produce pluripotent stem cell lines." The Institute, Stanford insisted, would be creating only cells, not human embryos. Leon Kass, the Council's chairman, protested. Stanford, he charged, had "decided to proceed with cloning research without public scrutiny and deliberation," and had "hurt the cause of public understanding" (Vaughan and Cool, 2003). When Proposition 71 hit the voting arena, it was clear that its authors had followed Stanford's lead, selecting terminology carefully to conceal the potential for developing human cloning technology. (During the campaign, one frequently replayed TV ad featured Irving Weissman, dressed in a white lab coat, asserting, "The chances for diseases to be cured by stem cell research are high, but only if we start . . . [W]e can hope for a single treatment with the right stem cells to cure diseases every family has." He was described only as a Stanford cancer researcher and as "California Scientist of the Year for 2002," and viewers were not informed that Weissman was a founder of one of the nation's leading publicly traded biotech companies, Stem Cells, Inc. Its stock rose 51% on the NASDAQ the Monday after Governor Schwarzenegger endorsed Proposition 71. At the time, Weissman owned 1.7 million shares (Stevens, 2007).)

Early in 2004, a well-funded army of paid signature gatherers could be seen in front of coffee houses and supermarkets asking pedestrians if they supported research into cures for a wide range of serious diseases such as diabetes, cancer, heart disease, Alzheimer's disease, multiple sclerosis, HIV/ AIDS, Parkinson's disease, ALS, osteoporosis, spinal cord injuries, and many other devastating medical conditions (Beeson and Stevens, 2004). Later that year, when the text of Prop 71 appeared on the ballot, the only reference made to cloning was the statement that use of bond proceeds to fund human reproductive cloning would be prohibited and that no Prop 71 funds would be used for "research involving human reproductive cloning." Nowhere did it explain why such a disclaimer was thought necessary. The appearance of this singular denial seemed awkwardly out of place, a fleeting flash of a bizarre demurral to an unstated accusation. Nowhere did it explain that the phrase used to refer to fundable research, "somatic cell nuclear transfer" was, in fact, cloning itself. Doubtless a strategy to avoid inflaming anti-choice critics, the word "embryo" never appears at all, not even in the context of IVF. In that context, embryos are "surplus products of *in vitro* fertilization treatments." Left as it stood, voters would never know that cloning technology crouched at the initiative's core. But a coalescing group of women's health advocates and progressive-left critics would labor to lift this veil.

The progressive left generally supported assisted reproductive technologies (ART) and stem cell research. But many of them also were concerned that the new reproductive and genetic technologies were at least as likely to exacerbate existing inequalities as to alleviate them (e.g., Arditti et al., 1984; Corea, 1985; Levine, 2002). Years before the 2004 election, the feminist women's health organization, Our Bodies, Ourselves (OBOS) had been working to raise consciousness about gender, race, and class issues that were intensified or provoked by some of the new technologies. It was an undertaking for which OBOS was uniquely qualified. The Boston Women's Health Collective (founded in 1971) and its transformational book, *Our Bodies, Ourselves*, launched "a full blown health movement," relates historian Ruth Rosen, one that taught "many Americans – not only feminists – to view themselves as medical consumers, rather than as passive patients." Moreover, writes Rosen, "the women's health movement created a relatively rare opportunity for cross-class and interracial activism" (Rosen, 2000, p. 180). OBOS Executive Director Judy Norsigian began working with two civil society groups concerned with oversight of emerging technologies from a progressive perspective: Council for Responsible Genetics (CRG), founded in 1983 by scientists and academics, including this book's co-author, Stuart Newman, and the fledgling Center for Genetics and Society (CGS), founded in 2001 by independent scholar-activists, Richard Hayes and Marcy Darnovsky.

In the summer of 2001, OBOS issued a call to the U.S. Congress to pass "a strong, effective ban on using human cloning to create a human being" and "a moratorium of five years on the use of cloning to create human embryos for research purposes" (Our Bodies Ourselves, 2001). The petition implicitly accepted the conceptual distinction between reproductive and research cloning, but attempted to stall the latter as well as the former. Besides OBOS, its signatories numbered over a hundred scholars, environmentalists, women's health activists and organizations, including the National Women's Health Network and the National Latina Health Organization. Less than a year later, CGS rallied progressives to support a similar position in an "Open letter to U.S. Senators on Human Cloning and Eugenic Engineering" (CGS, 2002). The letter, signed by 140 advocates for human rights, the environment, and social justice, also called for a ban on reproductive cloning and for a moratorium on the creation of clonal embryos.

In July 2001, Norsigian and Newman testified before a congressional committee in favor of a bill proposed by Rep. Dave Weldon, an anti-abortion Republican, that would ban the cloning of human embryos. The bill passed out of the House of Representatives in August. A *Boston Globe* editorial by Norsigian and Newman explained why a woman's health activist and a prochoice biologist, both "political progressives and defenders of reproductive autonomy," would take a position that found them "on the same side of the issue as antichoice conservatives: "embryo cloning will compromise women's health, turn their eggs and wombs into commodities, compromise their reproductive autonomy, and, with virtual certainty, lead to the production of 'experimental' human beings . . . we are convinced that the line must be drawn here" (Norsigian and Newman, 2001).

Within days of their editorial, the *San Francisco Chronicle* reported on what they characterized as the "odd-couple pairing" of abortion rights advocates and GOP conservatives on the U.S. cloning debate. It quoted Norsigian who spoke plainly: "This may be the only issue on the face of the Earth we agree on." It also quoted CRG Chair, Claire Nader, sister of the former third-party presidential candidate, Ralph Nader. "This is not a matter of left or right, liberal and conservative," she asserted. "This is about people who want to draw the line versus those who want to rush ahead before we know what the risks are" (Abate, 2001). This strange-bedfellow pairing in 2001 set a precedent. Bioentrepreneurs ramping up to the 2004 Prop 71 campaign may or may not have expected the repeat of such a parry. As for activists on the left and right, whether and how another "odd-couple pairing" could be undertaken was far from certain. The national politics of abortion was rife with sharpening divisions, even for those joined by their opposition to Proposition 71 (cf. Hurlbut, 2017).

Bountiful funding, trading on anti-Bush sentiment in a Democratic state, and promising cures for an impressive list of life-threatening conditions made easy work of the job of signature gathering. On June 2, California's Secretary of State certified the proposition for inclusion in the November election. On June 16, fewer than five months prior to the election, the California Catholic Conference convened what it envisioned as the launching of a broad-based coalition to defeat the initiative. Self-described as the "Official Voice of the Catholic Community in California's Public Policy Arena," its leadership understood that Proposition 71 could not be defeated without pro-choice voters mobilizing against it along with "pro-life" constituents. Critics on the Left eyed the opportunity warily. Would it be dominated by abortion rights opponents rather than a genuine coalition? With no financial resources and few scant months to rally, concerned parties on the Left stumbled toward the only near costless option open for democratic participation in the initiative process: writing a statement for the voter's guide. Would it be possible to cooperate with interested parties on the Right who first presented the opportunity? (Beeson and Stevens, 2004).

A pro-choice medical sociologist from California State University East Bay decided to take a chance. Acquainted with CGS since its inception, Diane Beeson shared its concerns over the looming prospect of inheritable genetic modification (designer babies) that embryo cloning enabled. Moreover, she was highly attuned to the history of unintended consequences associated with recklessly exposing women to large doses of exogenous hormones – something Prop 71's embrace of embryo cloning threatened to escalate dramatically. Beginning in 1938, for example, physicians incautiously began prescribing DES to pregnant women, gambling on the (mistaken, as it turned out) belief that the synthetic hormone would prevent miscarriages (CDC Update, n.d.). It was not until 1971, after mounting medical evidence of an alarming rate of vaginal cancer suffered by daughters born to mothers given DES, grass-roots action, and congressional intervention that the FDA finally announced that DES was contraindicated in pregnancy.

Progressive critics viewed Prop 71 as fitting squarely within the history of playing fast and loose with women's health. Women were being targeted to assume the understudied risks associated with egg harvesting, only no one was saying as much. For over three decades, young women had donated eggs to help other women suffering from infertility to have a child, as well as for their own infertility treatment. During all this time, as critics viewed it, the infertility industry had inadequately addressed cases where egg harvesting went bad. Ovarian Hyperstimulation Syndrome was a significant problem, sometimes with life threatening consequences. Moreover, the industry had demonstrated no interest in undertaking controlled studies to ascertain the long-term effects of the large doses of powerful synthetic hormones that egg harvesting required (Beeson and Lippman, 2006).

In the meantime, women's health advocates and organizations with an ear to the ground, such as Our Bodies Ourselves, took seriously the accounts that reached them, including a few cases where women had died (AHB website, n.d.). Prop 71 seemed to be riding on the fertility industry's lack of due diligence. A few scientific researchers were poised to increase dramatically the numbers of young women called upon to donate eggs – and just assume the unknown long-term risks and downplayed short-term risks of doing so. (Even now, infertility clinics and egg donor agencies routinely state that "risks are less than 1%" (Appendix E, Cool Testimony).)

Diane Beeson, like OBOS's Judy Norsigian, felt that circumstances necessitated a single-issue alliance with those otherwise known to challenge reproductive freedom. She attended the California Catholic Conference meeting. Of the roughly two-dozen attendees, only two were pro-choice, Beeson and Rex Greene, MD, an administrator with Mills Health Center in San Mateo. She learned that the group had decided to call itself "Doctors, Patients & Taxpayers for Fiscal Responsibility (DPTFR)." Hoping this was the beginning of a broad alliance, Beeson agreed to have her name posted on its website.

Since no other group had registered with the Secretary of State to oppose the proposition, **DPTFR** had the prerogative of writing the argument against the proposition and the rebuttal to the argument in favor of it. The main argument against the proposition, it was agreed, would be an economic one (that $3 billion dollars – $6 billion including interest – was excessive, and would be highly damaging to the state's economy). Two prominent political conservatives would sign it. The rebuttal was to be the principal vehicle for making clear that opposition to the measure came from both pro-choice and anti-choice camps.

Beeson pulled together a contingent of eight pro-choice feminists to assist in crafting a rebuttal. For weeks thereafter, she functioned as fulcrum, balancing uneasy negotiations between Left and Right. Once underway, the effort seemed likely to unravel at any moment. The feminists found conservatives' early drafts unacceptable. The two sides were not meeting face to face. Every word was a source of contention. Moreover, paring down complexities for public consumption seemed futile. Space limitations made it impossible to explain clearly how embryo cloning was a gateway technology to inheritable genetic modification or how it widened the opening door to human–animal chimerism. It also was difficult to explain how research cloning constituted the same technology as cloning for reproduction, that no oversight to prevent dissemination of the technology was proposed, or to explain the threats to women's health implicit in creating a demand for human eggs. The meager 250-word limit required difficult choices about what would be left out. With time running out, conservatives ultimately made the decision to accept whatever the feminists would say in the rebuttal and sign it.

Finally, the fractured "team," the two halves of which (save Beeson) could not have picked each other out of a line-up, submitted the Rebuttal, two words under limit and just short of the deadline:

Stem Cell Research? YES! Human Embryo Cloning? NO!
Here are just some of the many problems with Proposition 71:
** It specifically supports "embryo cloning" research – also called "somatic cell nuclear transfer" – which poses risks to women and unique ethical problems. To provide scientists with eggs for embryo cloning, at least initially, thousands of women may be subjected to the substantial risks of high dose hormones and egg extraction procedures *just* for the purposes of research. In addition, the perfection of embryo cloning technology – even if initially for medical therapies only – will increase the likelihood that human clones will be produced.
** Why privilege this research over other important research and medical needs, especially given the limits on how much California can invest? Why not issue bonds for programs that ALREADY have proven their cost effectiveness? Embryo stem cell research in nonhuman animals has produced only limited results. More compelling evidence of its efficacy should be required before a large commitment of public resources to study it in humans.

** Proponents are manipulating those seeking cures, building false hopes with exaggerated claims, and creating a costly program without adequate oversight or accountability. Stem cell research *should* be supported, but not this way. And don't be fooled by those who say that the opponents of Proposition 71 are all opposed to abortion and embryo stem cell research. Many of us are pro-choice, do not oppose all embryo stem cell research, and still oppose this initiative. Vote "No" on Proposition 71.

Judy Norsigian, Francine Coeytaux, Founder of the Pacific Institute for Women's Health, and this book's co-author, Tina Stevens, agreed to be its signatories for filing with the Secretary of State's office and eventual publication in the *Official Voter Information Guide*. As unlikely as it may have been that many would read it, for the first time voters would at least have the possibility to learn that funding Prop 71 facilitated embryo cloning and disregarded women's health. But bioentrepreneurs poised to bring legal action, sought to quash the effort (Stevens, 2007).

Within days of filing the Rebuttal, Prop 71 supporters brought a lawsuit against it, and against the DPTFR's Argument and "50 Word Summary" as well. Their brief alleged that statements made by opponents to Prop 71 were "false and misleading." The named "Petitioners" bringing suit, Paul Berg, Robert N. Klein, and Larry Goldstein, were seeking a writ of mandate ordering the Secretary of State to block the information from appearing in the voter's guide (Memorandum, n.d.). Petitioners objections to the Rebuttal fell on three statements: the first: "Human Embryo Cloning? NO!"; the second: "the perfection of embryo cloning technology ... will increase the likelihood [sic] human clones will be produced" (they left out the phrase: "even if initially for medical therapies only"); and the third statement, "thousands of women will be subjected to substantial risks of high dose hormones and egg extraction procedures *just* for the purposes of research."

This legal action brought into bold relief the troubling dynamics of biotech promotion in the post *Chakrabarty*–Bayh-Dole world. Proponents' petition drew upon and legitimated terminology since the early 1990s for the purpose of obscuring the nature of controversial scientific research. Prop 71, the petitioners insisted, did not involve the cloning of a human embryo; it sought to fund "nuclear transfer" which, it asserted, "is the cloning of embryonic stem cells for medical therapies, not the cloning of a human embryo, which is human reproductive cloning." A Declaration submitted by this book's co-author, Stuart Newman, on behalf of the Rebuttal exposed for the court this historic effort at obfuscation:

** Until Stanford University decided in the last year to stop using the terms "embryo cloning" and "cloned embryo" to describe the technique of producing human embryos by nuclear transfer and the products of this technique, these were the terms used virtually exclusively by scientists for these items.

** The term "cloned embryos" is still the term of art in this field of research for the products of nuclear transfer. A Medline search using this phrase turned up 42 uses of this term in article titles or abstracts during 2003–2004. In 2003, Ian Wilmut, the first scientist to clone a mammal, published an editorial in the journal, *Cloning and Stem Cells*, titled, "Human Cells from Cloned Embryos in Research and Therapy."

** The assertion that the viable product of nuclear transfer is not an embryo is equivalent to the assertion that organisms that develop from these products, such as Dolly the sheep, are not animals.

** Cloned mammalian embryos, the products of nuclear transfer, if permitted to develop to full term, are very likely to give rise to biologically abnormal or very sick individuals. This has been used by some to suggest that these entities are not genuine embryos. Following this line of argument leads to the proposition that human fetuses affected by Tay Sachs disease or Down Syndrome are not genuine human fetuses or the children they give rise to are not genuine human beings.

** Whether or not a scientist or physician intends to implant a cluster of cells does not determine whether or not it is an embryo. If it is a cluster of liver cells, for example, the intention to implant it does not make it an embryo. Correspondingly, if it is a blastocyst capable of giving rise to embryo stem cells, the lack of intention to implant it does not cause it not to be an embryo.

** To believe that the material nature of a biological entity changes depending on the intention of the investigator is an example of magical thinking, which is antithetical to modern science.

In the end, the court rejected the proponents' petition. The Superior Court Judge demanded the change of only one word in the Rebuttal: "thousands of women *will* be subjected to substantial risks" had to be changed to "thousands of women *may* be subjected to substantial risks." The attorney representing Respondents (the Rebuttal) characterized the Court's decision as a "slap in the face" to the petitioners (Beeson and Stevens, 2004).

Ironically, this slap in the face of Prop 71 proponents also proved an advantageous kick in their pants. A campaign consultant for the NO campaign recognized how even though critics may have won in the courts, proponents could rebound with strategic campaign advantages: "[T]he final result was clearly a major victory for our side," he explained,

and required that proponents open an entirely new defensive line to deal with arguments they had not anticipated and which clearly made their attempts to characterize Prop 71 opponents as "the usual suspects" a difficult sell . . .. Our early success forced them to begin the air war much earlier than they initially intended, and to spend more than they had planned.

(Beeson and Stevens, 2004, p. 42)

In terms of the cultural politics of biotechnology, the significance of the litigation was not its David's defeat of Goliath. The brief rebuttal however hard won was, after all, no match for the multimillion dollar media campaign that succeeded in propelling Prop 71 into law in the November election. The rebuttal's significance derives instead from the effort of bioentrepreneurs to suppress it. Reflecting on the trio of named petitioners challenging the rebuttal illuminates the cultural politics of bioentrepreneurialism.

Petitioner Robert Klein was the initiative's chief architect. A Stanford Law graduate and a prominent real estate developer, stem cell politics was for Klein, as one journalist noted in 2005, "intensely personal." His mother had been long suffering from Alzheimer's disease and his son had been diagnosed with juvenile diabetes. In 2005 he was quoted asking rhetorically, "How many chances in a lifetime do you have to impact human suffering in a really fundamental way, including possibly even in your own family?" (Hall, 2005). Since 2001, Klein had been an active fundraiser for the Juvenile Diabetes Research Foundation, one of the many patient advocacy groups enlisted to endorse Prop 71. Once the initiative passed, CIRM and the so-called Independent Citizens Oversight Committee (ICOC) devised to manage it were pelted by conflict of interest accusations, the grounds for which had been set in place early on. Klein, who gave $2.5 million as seed money, was one of Prop 71's largest contributors. He also was a significant donor to three of the four elected officials who, shortly after the election, unanimously nominated him to serve as chair of the ICOC board – a job description tailor-made by and for Klein (*Sacramento Bee* editorial, Dec 16, 2004). The ICOC position, which Klein held until 2011, and his leadership style after assuming office prompted one critic to refer to the institute as Klein's personal fiefdom (*California Stem Cell Report*, 2008). More commonly, he was referred to as the "stem cell czar" (*Sacramento Bee* editorial, 2004; Worth, 2011). (More on conflicts of interest is addressed in Chapter 4.)

Media and Prop 71 promotion described Petitioner Larry Goldstein as Professor at the University of California, San Diego or as a stem cell researcher. Typically unreported was that he also was a co-founder of the biotech company, Cytokinetics, Inc., and had functioned as lobbyist for the American Society for Cell Biology. The year after Prop 71 passed, Goldstein offered comment contradicting positions he and fellow Petitioners put forward in their 2004 legal memoranda.

After the election, Democratic State Senator Deborah Ortiz, an early champion of Prop 71, grew skeptical. She joined with Republican Senator George Runner to introduce bipartisan legislation seeking a three-year moratorium on egg harvesting procedures for the purpose of cloning research. The *Sacramento Bee* coverage of the proposed legislation, a measure meant to protect women's health, quoted Goldstein opposing it. Such a moratorium would, he said, have a "chilling effect and be very damaging for the research" (Mecoy, 2005). But opposing the moratorium on these grounds ran counter to arguments made in the 2004 legal action he and others brought to halt the Rebuttal. The petitioners argued in 2004 that Prop 71 funding would not result in thousands of women being subjected to the risks of high dose hormones and egg extraction because there were other sources for embryos and eggs. Existing frozen embryos (created for IVF but not selected for implantation) could be used for research, they claimed, or eggs could be obtained from women who might donate them while undergoing other procedures (e.g., hysterectomies). But if research could move forward in this way, then a moratorium on egg donations would *not* have a chilling effect on the research, as had been claimed in the previous year's litigation. Either the Petitioners' reasoning was incorrect when they instigated the legal action or it was incorrect when attempting later to discourage a moratorium. Was the lack of consistency a careless mistake, or was it the result of a cavalier attitude toward women's health encouraged by counterveiling professional and commercial interests?

The most prominent of the three petitioners was Paul Berg (Friedberg, 2014). Media and promotional material identified him often as Nobel Laureate or Stanford Professor Emeritus. Berg was also a co-founder of DNAX Institute of Cellular and Molecular Biology. He and colleagues launched the company the year Congress passed the Bayh-Dole legislation. In 1982, after Schering-Plough Pharmaceuticals acquired the company, he was a board member. But this commercial history was not an identifier during the Prop 71 campaign. Berg himself framed his support for the measure as motivated by courage, not commercial interests. "I got into this because I really see a threat, particularly when it is based on ideology or religion," Berg explained. "Proposition 71 is our way of saying, 'we're not gonna take it any more.' We have the wherewithal, and the citizenry is backing us to move forward with this opportunity" (Connolly, 2004). Portrayed in this way, it was an epic struggle of Science throwing off censorious Religion. Eventually, Prop 71 boosters would find an additional way of parlaying Berg's credentials, fashioning an effective if less plucky promotional tool. A July 2004 piece in the San Francisco Chronicle set the tone. Biotech journalist David Ewing Duncan reached back to the 1970s, fetching a 45-year-old Berg from the historic moment when he halted one of his own investigations. His lab was set to transfer DNA from a mammal into bacteria, for the first time. For Duncan, Berg's decision was, "a reverse Frankenstein story" (Duncan, 2004). Couched as heroic drama, the tale served to support the article's title prompt: "Why a cautious scientist supports the stem cell research initiative." The account, however, did not include crucial context that places Berg's decision in quite a different light.

This involved the events leading to the 1974 moratorium on recombinant DNA research and the 1975 Asilomar meeting discussed in Chapter 1. At Cold Spring Harbor Laboratory on Long Island, New York in 1971, particularly heated exchanges focused on projects said to be gearing up to use DNA cloning to increase the concentration of tumor-causing genes isolated from an infectious monkey virus, SV40, by a factor of an estimated hundred million. This was the virus Berg's lab was using. In his 1982 account, *Genetic Alchemy: The Social History of the Recombinant DNA Controversy*, Sheldon Krimsky relates how cell biologist Robert Pollack found that prospect troubling. What, he questioned, are the safety implications of such an increase? Would it overwhelm normal defense mechanisms? What Pollack learned more specifically from a Berg lab graduate student attending a course at Cold Spring Harbor that June left him dumbfounded.

Janet Mertz related how the Berg lab was nearing the point of isolating mammalian DNA, replicating it in *E. coli*, and using the cloned copies for a variety of experiments. ("Cloned" here is used for molecular replicas, not the organismal ones generated by SCNT.) The revelation caused a furor. Alarm rang out over the potential of a tumor virus genome growing in bacteria known to exist in people, escaping and colonizing in human beings. Pollack telephoned Berg. The experiment should not be done, he told him. Annoyed, Berg raised counter arguments. Only after additional strong negative feedback did Berg agree, several months later, to postpone moving forward.

Reflecting subsequently on the events, Pollack shared a significant realization: it had been easier for Pollack to recognize hazards and "raise noise about it" because he himself had not been working with the virus. "[A] shade comes over your eyes when the problems affect your own work," he explained. Had he been working with the virus, he asserted, he likely would have found his "own rationalization for not worrying about it" (Krimsky, 1985, p. 30). Pollack's astute observation raises crucial questions for the post *Chakrabarty*–Bayh-Dole era. Was Berg's decision to halt his project a tale of successful self-regulation, as Prop 71 promotion had spun it, or is it more properly understood as the result of urgings by knowledgeable but disinterested outsiders? The question is a crucial one in the age of bioentrepreneurialism. When patenting, trade secrets, intense competition, and profit motivation make information sharing a quaint, and ill-advised professional custom are there uninvolved outsiders with insider knowledge willing to challenge colleagues and risk professional censure? With vast amounts of money at stake, under what sorts of cultural constraints do scientists function who have careers of their own to protect? And when a

researcher *does* call for a halt, can it constitute anything more than a suggestion for a temporary time-out? (see Chapter 6). Questions such as these,
however, would not find public airing during the Prop 71 campaign.

The legal action against the prochoice Rebuttal underscores enduring dynamics of the cultural politics of bioentrepreneurialism: conflicts of
interest, burying or underplaying risks (in this case risks associated with
harvesting women's eggs), failing to address long-term implications (e.g.,
cloning and further human experimentation), and the way scientists can
resort to political-style branding. These dynamics remained operative
throughout the Prop 71 campaign. They helped fuel the continued association of the Rebuttal's drafters who remained in contact with each other after
the litigation's conclusion in early August. Stevens began a petition featur-
ing prochoice reasons to oppose the proposition and with Norsigian's help
and others grew the list of signatories. Unlike Prop 71 advocates, the emerging group's signature gatherers were unpaid. They also collected documents
for a website that Beeson paid for out of pocket. Eventually, they decided
to give their continuing association a name. The appellation they chose,
ProChoice Alliance Against Prop 71 (PCA), was inspired by the incredulous
reaction of journalists upon learning of the existence of prochoice critics.
A *Sacramento Bee* reporter challenged Beeson's pro-choice credentials.
"I would appreciate knowing those (prochoice) names," she emailed.
"Without a substantial number, it's hard to credibly say this is supported
by pro-choice [C]alifornians, especially when the other side is endorsed
by so many pro-choice organizations." Other prochoice critics had simi-
lar encounters with the press. Francine Coeytaux, founder of the Pacific
Institute for Women's Health (and a member of the earlier California
Cloning Commission), Southern California attorney Debra Greenfield, and
Loyola Law School faculty member Vicki Michel, spoke with members of the
LA Times editorial board on September 14. Greenfield described how,

> despite consistent affirmations, we were cynically and repeatedly
> questioned as to whether we were really pro-choice, and in favor of
> stem cell research. We insisted that our opposition was to the legal
> and ethical concerns with the proposed law itself, indeed what is on
> the ballot, not to the science itself.
>
> (Beeson and Stevens, 2004)

For those driving the emerging coalition, the encounters underscored the
importance of having a clear prochoice identifier.

The PCA website launched on September 26. With just weeks before
the election and constrained by competing work commitments and lack
of funds, they nevertheless managed to increase the list of supporters to 86
national and international individual signatories and eight organizational

signatories: The California Nurses Association; Center for Genetics and Society (CGS); Committee on Women, Population & the Environment; Council for Responsible Genetics; National Women's Health Network; Our Bodies Ourselves; The California Black Women's Health Project; and the Saheli Women's Resource Center. Eventually, CGS accepted management of the website (which until then had been a 15-year-old boy's project after homework) and, putting up some of its own funds, increased the effectiveness of the PCA significantly.

Few high-profile proponents would engage pro-choice critics. Pro-choice opponents to Prop 71 often found themselves with public engagements at which the initiative's supporters either could not be found or for which they cancelled out at the last minute. This happened repeatedly at radio, television, and public meetings sponsored by local League of Women Voters organizations (Beeson and Stevens, 2004). Bioentrepreneurs and their surrogates were manifestly more comfortable deploying familiar arguments aimed at the religious right. And while the PCA and its allies struggled to alert the public that Prop 71 facilitated human cloning technology and of the clinically removed basic science of embryonic stem cell research, the YES campaign permeated the media. Airwaves were chockablock with Nobel Laureates, scientists, and movie stars holding out the promise of imminent cures.

University of Alberta science and health policy instructor Timothy Caulfield detected hyperbolic language concerning stem cells in the popular press starting in the late 1990s. "The invisible hand of hype," as he termed it, brandished promises: "This is going to save lives. This is going to cure a variety of diseases" (CSC Report, June 23, 2016). This lure of hope intensified during the 2004 Prop 71 campaign. Among the more affecting of the Hollywood endorsements were actors Christopher Reeve, left paralyzed by a horseback riding accident, and Michael J. Fox, afflicted as a young man with Parkinson's disease. In 2012, *LA Times* columnist Michael Hiltzick characterized Prop 71 as having been "sold to a gullible public via candy-coated images of Christopher Reeve walking again and Michael J. Fox cured of Parkinson's." CIRM President Alan Trounson and Jonathan Thomas, Chairman of the agency's board demurred: "[N]o ads for Proposition 71 promised miraculous cures. They promised good science." Hiltzick was having none of it. The campaign had, he said,

> made a direct link between the creation of the $3-billion stem cell program and the discovery of definitive cures for specific diseases. It's true that science takes time and patience, but you wouldn't have known that from the Proposition 71 ad campaign.
>
> (Hiltzick, 2012)

When the smoke cleared over the battlegrounds of the 2004 election, California was in possession of a state-sanctioned right to undertake human embryonic stem cell research. The state was also now in possession of an unprecedented type of institution. The California Institute of Regenerative Medicine became the nation's first state agency dedicated to medical research with a mandate to find cures through the use of stem cells. It was also a state agency completely outside the normal channels of governance, and for which the state had no oversight authority.

Coda

Nationally, the politics of stem cell research and application stretches along a dynamic spectrum. Hype and hawking of hoped-for cures issue excitedly, at one end. At the other, steady hypothesis-testing pulses. Two 2017 reports characterize the spread's contrasting poles. In March, *The New England Journal of Medicine* reported on three women who sustained permanent and severe vision loss after receiving unproven stem cell "treatments" to control macular degeneration (Grady, March 2017). Owing to these and similar reports of premature application of stem cell "cures," the U.S. Food and Drug Administration in August "cracked down" on stem cell clinics it considered dangerous. FDA Commissioner Scott Gottlieb wanted, as the *New York Times* characterized, to protect the promise of stem cell therapies by weeding out fraudulent, unsafe players. As Gottlieb viewed it, "Products that are reliably and carefully developed will be harder to advance if bad actors are able to make hollow claims and market unsafe science" (Kaplan and Grady, 2017). By contrast, the *European Pharmaceutical Review* reported in November on two early clinical trials that seemed to suggest that embryonic stem cells might form a treatment for the dry form of macular degeneration (*EPR News*, November 15, 2017). As CIRM seeks to extend its life by securing funds, possibly once again from taxpayers, there lies a question just outside the scope of this book: where, along the hype hawking–hypothesis testing spectrum will CIRM oscillate as it balances inflamed expectation against patient safety? An answer must be left to future accounts.[3] The next chapter, "California Cloning: The Aftermath," picks up after passage of Prop 71 created CIRM. It recounts the institution's early conflicts of interests, its shifting posture toward the increased demand for women's eggs, and its consideration of emerging technologies involving eggs and embryos.

Notes

1 Portions of Chapters 3 and 4 were adapted from Beeson-Stevens, *Big Biotech and Abortion Politics* (2004).
2 Rossing, M.A., J.R. Daling, N.S. Weiss, D.E. Moore, and S.G. Self, "Ovarian Tumors in a Cohort of Infertile Women," *New England Journal of Medicine* 331 (1994): 771–776; van Leeuwen, F.E., H. Klip, T.M. Mooij, A.M.G. van de Swaluw,

C.B. Lambalk, M. Kortman, J.S.E. Laven, C.A.M. Jansen, F.M. Helmerhorst, B.J. Cohlen, et al. "Risk of Borderline and Invasive Ovarian Tumours after Ovarian Stimulation for *In Vitro* Fertilization in a Large Dutch Cohort," *Human Reproduction* 26 (2011): 3456–3465; Jensen, A., H. Sharif, K. Frederiksen, S. Krüger Kjær, "Use of Fertility Drugs and Risk of Ovarian Cancer: Danish Population Based Cohort Study," *BMJ* 338 (2009): b249; Li, L.L., J. Zhou, X.J. Qian, and Y.D. Chen, "Meta-Analysis on the Possible Association Between In Vitro Fertilization and Cancer Risk," *International Journal of Gynecological Cancer* 23 (2013): 16–24; Siristatidis, C., T.N. Sergentanis, P. Kanavidis, M. Trivella, M. Sotiraki, T. Mavromatis, A. Psaltopoulou, E. Skalkidou, and E.T. Petridou, "Controlled Ovarian Hyperstimulation for IVF: Impact on Ovarian, Endometrial and Cervical Cancer – A Systematic Review and Meta-Analysis," *Human Reproduction Update* 19 (2012): 105–123.

3 After this manuscript went to production, reportage by the *San Francisco Chronicle* assessed CIRM's track record and the agency's growing impetus to return to voters in 2020 seeking additional billions. CIRM appears to be recasting the agency's *raison d'être*. In 2004, supporters justified the creation of CIRM as a state solution to restrictive federal funding. Now, CIRM appears to be characterizing itself as a locally controlled pool of research money. Moreover, it does not seem to be exhibiting any more balanced an assessment of the research it funds than did 2004 boasts of research that it would fund. Current CIRM President, Maria Millan, proclaimed hyperbolically that, "Every single project we have is spectacular." Former Board Chair, Robert Klein, planning to lead the 2020 return to voters, sounded a familiar PR strategy: "Do you want your son to die? Are you going to wait? Is that the price you are prepared to pay?" (Allday and Palomino, September 6, 2018). In addition, the list of what CIRM claims as CIRM-funded cures are questionably related to its original raison d'etre under closer inspection. The cover of its 2017 annual report, for example, features a child cured of "bubble boy" syndrome (CIRM, 2017). But the technique used is a variation on the decades-old adult stem cell procedure (also used in the treatment of another case of the same condition, highly publicized by CIRM in its 2016 annual report) (see Chapter 4).

Sources Consulted for Chapter 3

Abate, Tom, "Odd-Couple Pairing in U.S. Cloning Debate / Abortion-Rights Activists Join GOP Conservatives," *SF Gate*, August 9, 2001: www.sfgate.com/news/article/Odd-couple-pairing-in-U-S-cloning-debate-2892207.php

Adams, Amy, "Prop 71 Victory Opens New Era for Research," *Stanford Report*, November 3, 2004: http://news.stanford.edu/news/2004/november3/med-proposition-1103.html

Allday, Erin, and Joaquin Palomino, "Lofty Promises, Limited Results: After 14 Years and $3 Billion Dollars, Has California's Bet on Stem Cells Paid Off?", *San Francisco Chronicle*, September 6, 2018.

Arditti, Rita, Renate Dueli Klein, and Shelley Minden Shelley, *Test Tube Women: What Future for Motherhood*, Thorsons edition, 1984; Pandora Press, 1989.

BallotPedia, https://ballotpedia.org/California_Proposition_71,_Stem_Cell_Research_(2004)

Beeson, Diane, and Abby Lippman, "Egg Harvesting for Stem Cell Research: Medical Risks and Ethical Problems," *Reproductive BioMedicine Online* 13(4) (2006): 573–579: http://static1.squarespace.com/static/55563925e4b08c2f72677221/t/557883f6e4b0bc4739ff508e/1433961462887/RBMOnline-Eggharvestingforstemcellres....pdf

Beeson, Diane, and M.L. Tina Stevens, "Big Biotech and Abortion Politics: The Progressive Campaign Against California's 2004 Stem Cell Initiative," 2004, available at www.humanebiotech.org

California Codes. Health and Safety Code Section 24185c(3)

California Stem Cell Report, December 7, 2008: http://californiastemcellreport.blogspot.com/2008/12/klein-says-he-cant-afford-to-work-for.html

California Stem Cell Report, "Stem Cells, Long Odds, and the Invisible 'Hand of Hype'," June 23, 2016: http://californiastemcellreport.blogspot.com/2016/06/stem-cells-long-odds-and-invisible-hand.html

California Stem Cell Report, "Stem Cells, Anonymous Commentary, and Silence Dogood," October 25, 2017: http://californiastemcellreport.blogspot.com/2017/10/stem-cells-anonymous-commentary-and.html?

CDC Update: www.cdc.gov/des/consumers/index.html

Center for Genetics and Society, "Open Letter to U.S. Senate on Human Cloning," March 19, 2002: www.geneticsandsociety.org/article.php?id=1958

Center for Genetics and Society, Campaign Website opposing Prop 71, November 10, 2004: www.geneticsandsociety.org/article.php?id=324

CIRM 2017 Annual Report: www.cirm.ca.gov/about-cirm/2017-annual-report

Cohen, Jon, "Stem Cell Pioneers," *Smithsonian Magazine*, December 2005: www.smithsonianmag.com/science-nature/stem-cell-pioneers-29038768/?no-ist

Colliver, Victoria, "Stem-Cell Agency Priced Out of SF, Moving to Oakland," *SFGate*, August 17, 2015: www.sfgate.com/health/article/Stem-cell-agency-priced-out-of-SF-moving-to-6449840.php?cmpid=email-desktop

Connolly, Ceci, "California Puts Stem Cells to a Popular Test," *The Washington Post*, October 25, 2004: www.washingtonpost.com/wp-dyn/articles/A59696-2004Oct24.html

Corea, Gena, *The Mother Machine: Reproductive Technologies from Artificial Insemination to Artificial Wombs*, London: The Women's Press, 1985.

Duncan, David Ewing, "Why a Cautious Scientist Supports the Stem Cell Research Initiative," *SF Gate*, July 25, 2004: www.sfgate.com/opinion/article/Why-a-cautious-scientist-supports-the-stem-cell-2705835.php

EPR News, "Stem Cell Therapy Shows Promise for Macular Degeneration," *European Pharmaceutical Review*, November 15, 2017: www.europeanpharmaceuticalreview.com/news/69332/stem-cell-therapy-macular-degeneration/

Friedberg, Errol C., *A Biography of Paul Berg: The Recombinant DNA Controversy Revisited*, Hackensack, NJ: World Scientific Publishing, 2014.

Grady, Denise, "Patients Lose Sight After Stem Cells Are Injected Into Their Eyes," *The New York Times*, March 15, 2017: www.nytimes.com/2017/03/15/health/eyes-stem-cells-injections.html

Greenfield, Debra. "Impatient Proponents: What's Wrong with the California Stem Cell Research and Cures Act," *Hasting's Center Report*, Sept/Oct 2004: 33–35.

Hall, Carl T., "STEM CELLS: The $3 Billion Bet / One Man's Scientific Mission / Housing Developer Leads California's Research Effort," *SF Gate*, April 11, 2005: www.sfgate.com/news/article/STEM-CELLS-The-3-Billion-Bet-One-man-s-2642279.php#photo-2121970

Hiltzick, Michael, "Did the California Stem Cell Program Promise Miracle Cures?" *LA Times*, May 30, 2012: http://articles.latimes.com/2012/may/30/news/la-mo-stem-cell-20120530

Hurlbut, Benjamin J., *Experiments in Democracy: Human Embryo Cloning and the Politics of Bioethics*, New York: Columbia University Press, 2017.

"Is Human Cloning Here? Chinese Scientist Ready to Clone People at His 'Replication Factory'," *RT*, December 1, 2015: www.rt.com/news/324110-china-human-cloning-factory/

Jensen, David, "California's Stem Cell Agency Will Run Out of MONEY in Three Years. Should Voters OK Spending More?" *Sacramento Bee*, January 17, 2017: www.sacbee.com/news/local/health-and-medicine/article126899909.html

Jewett, Christina, "Women Fear Drug They Used to Halt Puberty Led to Health Problems," *California Healthline*, February 2, 2017: http://californiahealthline.org/news/women-fear-drug-they-used-to-halt-puberty-led-to-health-problems/

Kaplan, Sheila, and Denise Grady, "F.D.A. Cracks Down on 'Unscrupulous' Stem Cell Clinics," *New York Times*, August 28, 2017: www.nytimes.com/2017/08/28/health/fda-stem-cell.html

Krantz, Matt, "Stem Cells Not Drawing Cash," *USA Today*, September 6, 2001: http://usatoday30.usatoday.com/news/health/2001-09-06-stem-cells-investment.htm#more

Krimsky, Sheldon, *Genetic Alchemy: The Social History of the Recombinant DNA Controversy*, Cambridge, MA: MIT Press, 1985.

Levine, Judith, "What Human Genetic Modification Means for Women," World Watch, July/August 2002, pp. 26–29; and in Jeremy Gruber, and Sheldon Krimsky, *Biotechnology in Our Lives: What Modern Genetics Can Tell You About Assisted Reproduction, Human Behavior, Personalized Medicine, and Much More*, New York: Skyhorse Publishing, June 2013.

McLean, Margaret R., "Red Light, Green Light: The California Cloning and Stem Cell Laws," *Hastings Center Report* 32(6) (2002): 7.

Mecoy, Laura, "Stem Cell Allies Divided Over Egg Collection," *The Sacramento Bee*, March 27, 2005, archived by Center for Genetics and Society: www.geneticsandsociety.org/article.php?id=1615

Memorandum of Points and Authorities in Support of Petition for Writ of Mandate and Alternative Writ of Mandate/Order to Show Cause (n.d., no case number), Paul Berg, PhD; Robert Klein; and Larry Goldstein, Petitioners vs. Kelvin Shelly, Secretary of State of California, Respondent, Geoff Brandt, State Printer; Bill Lockyer. Attorney General of California; Tom McClintock; H. Rex Green; John M. W. Moorlach; Judy Norsigian; Francine Coeytaux; Tina Stevens; Does I through X, inclusive, Real Parties In Interest. (Eventually, Case No. 04CS01015, Hearing Date, August 4, 2004.)

Murugan, Varnee, "Embryonic Stem Cell Research: A Decade of Debate from Bush to Obama," *Yale Journal of Biology and Medicine* 82(3) (2009): 101–103: www.ncbi.nlm.nih.gov/pmc/articles/PMC2744932/

Newman, Stuart A., Declaration of Dr. Stuart A. Newman, PhD in Opposition to Petition for Writ of Mandate and Alternative Writ of Mandate/Order to Show Cause, Case No. 04CS01015, August 4, 2004.

Norsigian, Judy, and Stuart Newman, "Ban Human Cloning Right Now," *Boston Globe*, August 3, 2001, online at: www.geneticsandsociety.org/article.php?id=169

Norsigian, Judy, "Stem Cell Research and Embryo Cloning: Involving Laypersons in the Public Debates," *New England Law Review* 39 (2005): 701–708.

Novak, Kris, "New Stanford Institute Sparks Cloning Quarrel," *Nature Medicine* 9(2) (February 2003): 156–157.

Official Voter Information Guide, California General Election, Text of Proposed Laws, Proposition 71: http://vig.cdn.sos.ca.gov/2004/general/propositions/prop71text.pdf; http://vig.cdn.sos.ca.gov/2004/general/english.pdf

Our Bodies Ourselves, "Statement on Human Cloning," June 2001: https://web.archive.org/web/20140507161703/http://www.ourbodiesourselves.org/book/companion.asp?id=31&compID=67&page=2

Priest, C.A., N.C. Manley, J. Denham, E.D. Wirth, 3rd, and J.S. Lebkowski, "Preclinical Safety of Human Embryonic Stem Cell-Derived Oligodendrocyte Progenitors Supporting Clinical Trials in Spinal Cord Injury," *Regen Med* 10 (2015): 939–958.

Quintero, Maria S., "Cloning Californians: Report of the California Advisory Committee on Human Cloning and Recent Cloning-Related Legislation," *Santa Clara High Tech. L.J.* 18(417) (2001): hp://digitalcommons.law.scu.edu/chtlj/vol18/iss3/8

Regaldo, Antonio, and Michael Waldholz, "Ballot Drive Puts Stem Cell Funding in Voter's Hands," *Wall Street Journal*, March 31, 2003.

Reynolds, Jesse, blog post: "CIRM Won't Give Up on Eggs for Cloning Based Research," December 3, 2008: www.geneticsandsociety.org/biopolitical-times/cirm-wont-give-eggs-cloning-based-work

Rosen, Ruth, *The World Split Open: How the Modern Women's Movement Changed America*, New York: Penguin Books, 2000.

Sacramento Bee, Editorial, "Stem Cell Czar?," December 16, 2004: www.geneticsandsociety.org/article.php?id=1553

Servick, Kelly, "Failed Spinal Cord Trial Offers Cautionary Tale," *Science* 355(6326) (February 17, 2017).

Smith, Rebecca, "'I Can Create Neanderthal Baby, I Just Need Willing Woman'," *The Telegraph* (January 20, 2013): www.telegraph.co.uk/news/science/9814620/I-can-create-Neanderthal-baby-I-just-need-willing-woman.html

"StemCells, Inc. Appoints Dr. Alan Trounson to Board of Directors," *The Street*, July 7, 2014: www.thestreet.com/story/12765616/1/stemcells-inc-appoints-dr-alan-trounson-to-board-of-directors.html

Stevens, M.L. Tina, "Intellectual Capital and Voting Booth Bioethics," Lisa Eckenwiler and Felicia G. Cohn, eds, *The Ethics of Bioethics: Mapping the Moral Landscape*. Baltimore, MD: Johns Hopkins University Press, June 2007, pp. 59–73.

Thompson, Charis, *Good Science: Ethical Choreography of Stem Cell Research*, Cambridge, MA: MIT Press, December 20, 2013.

Vaughan, Christopher, and Kevin Cool, "Cell Division," *Stanford Alumni*, May/June 2003: https://alumni.stanford.edu/get/page/magazine/article/?article_id=36663

"Voter Information Guide," California General Election, November 2, 2004, Proposition 71 text, pp. 147–155; Analysis by Legislative Analyst, pp. 68–71; Argument in Favor of Proposition 71, Rebuttal to Argument of Proposition 71, Argument Against Proposition 71, Rebuttal to Argument Against Proposition 71, pp. 72–73. http://vig.cdn.sos.ca.gov/2004/general/english.pdf

Witherspoon Council, "The Stem Cell Debates: Lessons for Science and Politics," *The New Atlantis: A Journal of Technology and Science* (Winter 2012): www.thenewatlantis.com/publications/the-stem-cell-debates-lessons-for-science-and-politics

Wobus, A.M., and K.R. Boheler, "Embryonic Stem Cells: Prospects for Developmental Biology and Cell Therapy," *Physiol Rev* 85 (2005): 635–678.

Worth, Katie, "California Stem Cell Czar Offered Top Pay, Asked to Name Job Description," *San Francisco Examiner*, May 29, 2011: www.sfexaminer.com/california-stem-cell-czar-offered-top-pay-asked-to-name-job-description/

Yes on 71 archived website: http://digital.library.ucla.edu/websites/2004_996_027/index.htm; http://digital.library.ucla.edu/websites/2004_996_027/quotes.php.htm; http://digital.library.ucla.edu/websites/2004_996_027/tv_radio.php.htm; http://digital.library.ucla.edu/websites/2004_996_027/documents/KeyTalkingPoints.pdf; http://digital.library.ucla.edu/websites/2004_996_027/news_clip_0824_update.php.htm

4
California Cloning
The Aftermath

They make proposals to themselves . . . regarding what should be funded.
(Harold Shapiro, Institute of Medicine,
quoted in Jensen, 2016)

Prop 71 was the product of a juggernaut. The elements of that unstoppable force could be identified clearly in retrospect. Failing first in the legislature, bioentrepreneurs targeted California's initiative process to carry the measure into law. As discussed in Chapter 3, strategies that keyed the measure for success included: the adoption of terminology designed to obscure the measure's reliance on cloning technology; ignoring how cloning technology advances human genetic engineering; burying the need for women's eggs and its associated health risks; claiming, without sound reasoning, that funding Prop 71 would be a tonic for a state suffering an economic recession; spending millions saturating media with promises of cures, offering no clear scientific evidence supporting avowals that it was the best route to cures; litigating to silence prochoice critics and refusing to publicly engage progressive critics. Nullifying controversy by refusing to recognize it proved an effective strategy.

The measure, now law, created a constitutional right to conduct stem cell research in California including – though the text obscured it – a right to create human clonal embryos. This guarantee came with a $3 billion program of publicly financed research ($6 billion total cost to taxpayers after interest) to be administered by Prop 71's new bureaucratic incarnation, the California Institute of Regenerative Medicine (CIRM). CIRM's governing board, populated by industry insiders and patient advocates, bore scant resemblance to the impartial body evoked by its self-chosen title, the Independent Citizens Oversight Committee (ICOC). The provisions of the ballot measure shielded CIRM and its activities from legislative scrutiny for three years, and permanently closed it off from broader popular input. Even after three years, any amendment would have to pass by 70% of both houses and have the approval of the Governor, an almost impossible bar to any intervention. And, in spite of campaign promises to the contrary, the agency was not required to return to the state any of the private profits this public investment might generate. In 2010, Senate Bill 1064 passed into law, which made some changes designed to improve CIRM's transparency and accountability, and regulating intellectual property rights (SB 1064). Yet, concerns about transparency, accountability, and efficacy continue to stalk the agency.

After almost a decade and a half since Proposition 71 passed in 2004, how has the state's investment fared? Has it brought forth promised cures? Has it fortified the state's economy with "thousands of new jobs and millions in new state revenues"? (Voter Information Guide, 2004). Has its management garnered good governance approval? In February 2018, Marcy Darnovsky, Director of the watchdog, Center for Genetics and Society, concluded that CIRM needs to do better:

> Many Californians voted to establish CIRM because they believed the promises that its backers were making: that we'd soon see revolutionary medical breakthroughs, that our state would get back a billion dollars or more in royalties, that the agency would be run by an 'independent' board. Almost a decade and a half later, none of that has come to pass. The rules and regulations about royalty returns to California are confusing and unclear, and need to be made far more transparent.
>
> (CSCR, February 14, 2018)

As of this writing, CIRM is expected to run out of money in three years' time (Jensen, 2017). How and whether it will be re-funded is a matter of some speculation. A return to voters requesting a re-up of bond financing, this time for $5 billion, was one suggestion (Rodota and Munos, 2017). Priming to loosen purse strings, the cover of CIRM's 2016 Annual Report featured an image that tugged unassailably at heart strings: six-year-old "Evangelina," playing joyously astride a rocking horse, pronounced cured of severe combined immunodeficiency disease or "bubble boy syndrome." Inside, five more breakout vignettes spotlighted individuals improved, if not cured, after treatments involving stem cells. How much CIRM can uniquely claim for itself from this handful of accounts is unclear. Only Evangelina's case constituted a cure. The treatment in her case, however, did not involve ES cells (embryonic stem cells), CIRM's *raison d'être*. Evangelina's own "adult" blood stem cells were used in the *ex vivo* procedure described in Chapter 1, i.e., her cure came out of a type of research that is eligible for federal funding and, in fact, was in use well before CIRM existed (Graze et al., 1979; Futterman and Lemberg, 2004). The remaining stories related improvements, not cures, of uncertain duration and extent, or involving adult, not ES cells. Twenty-seven clinical trials were underway at the time, some involving adult stem cells, some involving ES cells. For then CIRM President and CEO, C. Randall Mills, such a track record warranted continuing faith in the agency. A smiling photo framed beside his brief address, Mills' optimism echoed distant campaign promises resounding 12 years earlier: "we can now see cures and treatments on the horizon," he says. CIRM, he pledged, "will never stop looking for ways to do more, faster, with greater success" (CIRM Annual Report, 2016).

CIRM's critics weigh such familiar sounding lofty pledges against the history of its factual record. While cures failed to materialize, disputes over questionable expenditures and conflict of interest controversies swirled persistently. In January 2017, health and science investigative journalist Charles Piller characterized Prop 71's "unprecedented experiment in medical research by direct democracy," as a "response to federal funding limits for embryonic stem cell research . . . [that was] sold with a simple pitch: The money would rapidly yield cures for devastating human diseases . . .. That hasn't happened." He contrasted the "trickle" of clinical trials to the millions that went to building infrastructure, labs, and buildings at universities – some public, but also private, e.g., Stanford University and the University of Southern California. Public monies funding private institutions is a troubling practice in itself (Piller, 2017). CGS' Marcy Danovsky cut to the chase: "You could make an argument that California taxpayer money should go to build new facilities on state university campuses," she said. "But I don't see an argument for Stanford getting fancy new buildings from California taxpayer money" (Piller, 2017).

Conflicts of Interest in the Running of CIRM

In 2008, the leading science journal, *Nature*, warned of "cronyism" at the California Institute of Regenerative Medicine (Jensen, 2016). California's bipartisan independent good-government state oversight agency, the Little Hoover Commission, buttressed the concern in June 2009. The Commission told the State's legislature and governor that CIRM should restructure its governance structure, "to improve transparency and accountability and speed its success in finding cures through stem cell research" (Hoover Commission, 2009). In his 2008 testimony before the Hoover Commission, CGS's Jesse Reynolds connected CIRM's transparency and accountability deficits to its leadership's conflicts of interest:

> This lack of accountability and oversight has been most apparent in some actions of the CIRM's leadership, particular those of the chair of the ICOC [Robert Klein]. For example, when a key Senate supporter of the CIRM convened a joint informational session of the Senate and Assembly Health Committees, he rejected the invitation, saying "The Legislature is not needed." Since then, he has simultaneously served as the leader of both the CIRM and a series of private stem cell research lobbying organizations. If the ICOC chair were subject to

appropriate structures of accountability, some of his actions as head of these lobbying groups would have come under significant scrutiny and likely have been viewed as unacceptable for a public official.

(Reynolds, 2008a)

In 2012, the U.S. Institute of Medicine (IOM) heard similar critical testimony. Summarized in *Nature*'s newsblog: "The board is very large and vulnerable to charges of conflict of interest, because all of its members – from university officials to disease advocates – stand to benefit from the research that the agency is funding, the IOM panel heard." The way UC Berkeley School of Law corporate-governance scholar Ken Taymor saw it, CIRM's board was unable to review grant awards rigorously owing to its lack of transparency and independence. "Even to the extent that there is an opportunity for some real questioning that might occur about the validity of a grant or pursuing a certain program," Taymor said, "that just isn't done. There's too much incentive to [say], 'Well, I should support this so my disease or pet project will be supported,'" he explained (*Nature News Blog*, 2012). A spokesperson for the independent Santa Monica-based Consumer Watchdog organization waxed similarly critical, "The IOM's critical report echoes what every independent evaluator has said in the past. As we have repeated from the beginning, CIRM suffers from built-in conflicts of interest and needs to separate the board's oversight function from day-to-day management." The watchdog organization was not optimistic. "It's long past time to make the changes the report calls for, but given the spin the agency put on its response – saying the report praises the 'agency as a bold innovation' – shows it's business as usual" (Simpson, 2012). In 2013, the IOM review concluded its study, reporting that CIRM's conflicts of interest had raised questions concerning "the integrity and independence of some of CIRM's decisions" (Piller, 2017).

In May 2011, the *San Francisco Examiner* recounted how Robert Klein, CIRM's media-dubbed "stem cell czar," had to withdraw his nomination for his successor who was then too marred by allegations of backroom deals. Subsequently, the board failed to devise a consensus-supported job description for the position. It did, however, agree on a maximum full-time salary: $508,000, far higher than the salaries of other significant public employees, California's governor: $174,000; the Lieutenant Governor: $130,000; San Francisco's Mayor $248,000 (*San Francisco Examiner*, 2011).

CIRM awarded many of the largest grants to recipients with representatives on its governing board. By 2012, roughly 90% of CIRM's grants had gone to such institutions (Hiltzick, 2012). "Stanford, whose endowment is among the top five nationally, and USC," Piller reports, "have received more than $70 million for major building projects, and hundreds of millions more for labs and research. Stanford alone has been favored with $1 out of every $7 CIRM has approved."

During the Prop 71 campaign, commercials often featured Irving Wiessman, identifying him as a Stanford research professor. But Weissman also was a co-founder of StemCells Inc. In 2012 his company ranked first of all CIRM's corporate awardees (Hiltzick, 2012).

> He's also been a leading beneficiary of CIRM funding, listed as the principal researcher on three grants worth a total of $24.5 million. The agency also contributed $43.6 million toward the construction of his institute's glittering $200-million research building on the Stanford campus.
> (Hiltzick, 2012)

In November 2011, Biotech giant, Geron Inc. announced a move that resulted in swift criticism of CIRM: Geron was quitting the stem cell business. The company took pains to defend the promise of stem cells. The decision to leave the field, it claimed, was just a shift in support to more mature cancer therapies. But CIRM had lent $25 million to Geron for a clinical trial testing ESCs in spinal cord patients (Pollack, 2011).

LA Times' CIRM watchdog Hiltzick wrote:

> [T]here's evidence that CIRM, anxious to show progress toward bringing stem cell therapies to market, downplayed legitimate questions about the state of Geron's science and the design of the clinical trial . . . . Geron had been criticized in the past for over-promising results. Some researchers questioned the design of its clinical trial and even whether spinal cord injury was the best subject for the first tests of stem cell therapies on humans . . . CIRM didn't disclose in advance that Geron was the loan applicant. Nor did it disclose that its own scientific review panel had awarded the Geron trial a scientific score of only 66 out of 100 . . . five of the CIRM board's 28 members had to recuse themselves from the vote because of conflicts of interest. That underscores the structural weakness of the board, which is drawn almost entirely from patient advocacy groups and biomedical research institutions angling for grants.
> (Hiltzick, *LA Times*, December 7, 2011)

Stem cell researcher Robert Lanza of Advanced Cell Technology described Geron's move as, "very unfortunate for the field" (Pollack, 2011). Despite such opinions and criticism of CIRM's dicey grant-vetting standards, CIRM persisted the next year in funding the speculative research of one of its favored grantees, Irving Weissman's Stem Cells, Inc. The governing board approved $40 million of awards for the company in 2012. CIRM's scientific panels twice recommended rejecting one of the applications. Then, in his first appearance before the board since stepping down as CIRM's governing chair, Robert Klein led an appeal. The decisions to reject were overturned. Some board members complained that Klein's effort smacked of "lobbying," "politicking," and "friends" calling "friends." *LA Times* editorialist Hiltzick opined that, "the record suggests that the handling of the StemCells appeal was at best haphazard and at worst redolent of cronyism" (Hiltzick, *LA Times*, October 17, 2012).

A 2017 report in *Science* would come to underscore the corrosive implications of funding such poorly vetted research. The article, "Failed spinal cord trial offers cautionary tale," made clear that the stem cell lines produced by StemCells, Inc. had "lacked support from animal studies." A UC Irvine study found that the stem cell lines, "provided no benefits to mice with upper spinal cord injury." Moreover, according to *Science*, the UC Irvine researchers reported that "they showed these preliminary results to StemCells Inc. in July 2014 and were surprised when the company still launched its clinical trial." It quoted one of the researchers saying, "[t]here's no ethical way you could go forward with this trial in people" (Servick, 2017). By 2016 the investment community (too often the decisive arbiter of these activities), had spoken, and StemCells, Inc. was gone (Reuters, 2016).

Yet another embarrassment for CIRM set media outlets crackling in July 2014: "The Trounson Affair." Within a week of stepping down from CIRM leadership Alan Trounson, the Australian infertility specialist who served as CIRM president from 2008–2014, joined the board of StemCells, Inc. The company was a prime beneficiary of sometimes controversial CIRM awards, including in 2012 a $19.4 million forgivable loan for what the *San Francisco Business Times* characterized as, "a scientifically questionable preclinical program in Alzheimer's disease" (Leuty, 2014). CIRM itself acknowledged that "even the appearance of conflicts of interest" needed to be taken seriously. An *Orange County Register* editorial screamed that it was "Time to Target California Cronyism" (Orange County Register, 2014). Center for Genetics and Society associate Pete Shanks wrote, "Let's be blunt: This looks like a pay-off. Technically, what Trounson and Weissman and StemCells, Inc., just did may not be illegal. But it's shameless."

CIRM swiftly acknowledged not wrongdoing but the "risk" of it: "the appointment of CIRM's former president to the board of directors of a CIRM loan recipient creates a risk of a conflict of interest" and pledged a "full review" of the relationship between CIRM and StemCells, Inc. (CIRM, July 9, 2014). A mere 15 days later on July 24, the agency announced that its "limited" investigation of documents, emails, and interviews with staff found that Trounson had not attempted to influence decisions in favor of StemCells, Inc. (*California Stem Cell Report*, July 24, 2014). But two years later in 2016, former *Sacramento Bee* editor David Jensen learned that Stem Cells, Inc. gave Trounson $443,500 in stock and cash over a two year period. This reportage also noted that while Trounson was head of CIRM, the agency awarded StemCells, Inc., more than $40 million. Furthermore, "In 2014, when Trounson served less than six months on the StemCells, Inc. board, his total compensation was more than double that of any other board member." John Simpson of Santa Monica's Consumer Watchdog organization opined that,

> Trounson's joining the StemCells Inc. board a mere seven days after quitting as CIRM's president at the time smacked of being a payback for a job well done on behalf of the company when he should have been looking out for the taxpayers' money and interests. It was a blatant conflict that undermined the agency's credibility. Now we know his price tag.
>
> (Jensen, 2016)

C. Randal Mills, former president and CEO of Osiris Therapeutics, replaced Trounson as CIRM president, and is generally credited with improving CIRM's transparency and the efficiency of its grant making process. After serving only three years as CIRM president, Mills unexpectedly announced in 2017 that he planned to step down at the end of June. He strongly endorsed Maria Millan to assume CIRM's helm upon his departure (she has since done so). Notably, Millan had ties both to Stanford University, where she directed the Pediatric Liver and Kidney Transplant Program, and to StemCells, Inc., where she served as the company's vice president and acting chief medical officer (Fikes, 2017).

The history of CIRM's struggles with conflicts of interest chafes against its self-promoting narrative of success. Moreover, the agency's practices and policies engender two ethical quagmires: the increased pressure to acquire young women's eggs (only partly reduced by the introduction of new techniques such as induced pluripotent stem cells (iPSC, see Chapter 6) and the largely unexamined enabling relationship between ESC research and human genetic engineering. For women's health advocates watching developments unfold, the first item was the most urgent. The years following passage of Prop 71 inflamed the market for women's eggs. Previously, the infertility industry drove the trade in human ova. Now that industry rides in tandem with the ESC research community in seeking laws and setting policies to satisfy the swelling demand for oocytes. Much of the bioresearch on human embryos is conducted at major medical centers with fertility clinics. Indeed, the Oregon lab that in 2017 created the first gene edited human embryos accesses eggs from a fertility clinic just three flights down (Stein, 2017).

The Egg Wars

Proposition 71 had banned paying women to donate their eggs (Thompson, 2013, p. 107). But a 2007 report from the Ethics Committee of the American

Society for Reproductive Medicine only confirmed what critics had warned of when Prop 71 passed in 2004: "recent scientific developments suggest that oocyte donation may become an important process in the field of stem cell research" (Thompson, 2013, p. 105). A 2007 IOM report made the suggestion manifest. In 2006, CIRM requested the IOM to assess the medical risks associated with egg extraction. After holding a meeting in September, it released its "workshop report" in February 2007: "Assessing the Medical Risks of Human Oocyte Donation for Stem Cell Research." Its assessments recognized the inadequacy of research to date. "One of the most striking facts about in vitro fertilization," it stated, "is just how little is known with certainty about the long-term health outcomes for the women who undergo the procedure." It acknowledged that even "that limited knowledge is not directly applicable to the safety of ooctye donation for research." Good data collection and long-term studies were needed, it suggested. It was, it said, necessary to reduce risks to women. Then it offered an astonishingly unsupported conclusion: egg extraction was a remarkably safe procedure. The IOM's CIRM-requested report, gave CIRM a green light to increase the demand for women's eggs (Darnovsky, 2007; IOM, 2007). Always, as the activists opposing Prop 71 understood, there would be intensifying pressure to gain access to eggs, with insufficient attention paid to the health and welfare of the young women assuming all the risks (Parisian, 2005).

Following the election, they worked to forestall the erosion of what few ethical standards were in place and to press for increased transparency and equity of emerging practices. California's SB 1260 is an example of this effort. The Center for Genetics and Society, Planned Parenthood Affiliates of California, and the Pro-Choice Alliance for Responsible Research worked closely with then California Senator Deborah Ortiz (Democrat, Sacramento) to pass SB 1260. (Ortiz had been an early champion of stem cell research and a strong supporter of Prop 71.) The new law sought to close a gap whereby guidelines created by CIRM for publicly funded research would not apply to research privately funded. The law, when passed in 2007, ensured that, "women in California who provide eggs for private research be accorded all established federal and state protections for human research subjects" (CGS SB 1260), including prohibiting "paying for eggs for research beyond compensation for direct expenses" (AHB et al., 2013). Significantly, although the bill was limited to eggs provided for research and did not affect women who donate eggs for the fertility industry, it was opposed by the American Society for Reproductive Medicine (ASRM) (Darnovsky, 2007). ASRM's desire to establish and maintain a wide open market in eggs would be evidenced time and again in the ensuing years.

In 2009, the Alliance for Humane Biotechnology (AHB) (with which this book's authors are associated) worked with California Assembly Member Marty Block (D – San Diego) assisting him in introducing AB 1317.

The bill required that egg broker ads reference the existence of health risks associated with egg harvesting. Joining AHB, CGS, OBOS, and PCARR in supporting the bill were California NOW, Breast Cancer Fund, and the American Association of University Women. The law's supporters intended to even the informational playing field between egg donors and egg seekers. Egg brokers and fertility clinics routinely posted flyers on college campuses and placed ads in student newspapers offering female students thousands of dollars for their eggs. Until AB 1317, there was no requirement that such inducements be tempered in any way by acknowledgment of the existence of health risks.

The law became the first of its kind and set a precedent signaling the importance of establishing full disclosure before a truly informed consent is possible. But it did not, in itself, establish that full disclosure. This was in large measure owing to lobbying efforts on the part of ASRM which secured an exemption for its members. (During a legislative hearing on the bill, the lobbyist for ASRM asserted that the organization favored the idea of a registry. There is, however, no public evidence of ASRM seriously working toward establishing a mandatory, long-term, rigorously controlled, independent, donor health registry, then or now.) Moreover, instead of requiring strong language indicating that the long-term health risks associated with egg donation were unknown, ads need state only that, "As with any medical procedure, there may be risks associated with human egg donation." An additional clause required notice that, "Before an egg donor agrees to begin the egg donation process, and signs a legally binding contract, she is required to receive specific information on the known risks of egg donation." This references only known risks, thereby skirting the enormous problem that there are few studies and no health registries tracking long-term risks of egg harvesting.[1] (Enforcement of AB 1317 presents an additional limitation. For a while, AHB collected ads not in compliance and sent them to the California Attorney General's office. But lack of funds to continue such watchdog effort limited this strategy.)

Frustrated, activists instigated a "Petition for Human Egg Extraction Health Registry & for Warnings on Ads and Notices Seeking Egg Donors" in 2011[2] (Petition 2011, AHB, OBOS, et al.). Institutional support for the petition included Alliance for Humane Biotechnology, The California Nurses Association, Center for Genetics and Society, Council for Responsible Genetics, National Women's Health Network, National Organization for Women, Our Bodies Ourselves, and ProChoice Alliance for Responsible Research (Petition, AHB website). Results such as SB 1260, AB 1317, the egg donor petition and the tireless efforts of women such as Jennifer Schneider (see p. 85) and the founders of We Are Egg Donors (see p. 86) represent committed effort on the part of volunteer and scantly funded activists to create protections and improve informed consent for egg donors.

Given the complexities of the issues and the lack of dedicated time and money available to raise public awareness, such labors are impressive. Even so, they are no match for the fertility industry's and research community's well-funded persistent drive to widen access to human eggs regardless of long-term health risks to women.

A few years after the passage of Prop 71, CIRM president Alan Tounson and chairman of the board Robert Klein first brought to light CIRM's stealthy efforts to undo Proposition 71's prohibition on paying women for their eggs. At a CIRM Standards Working Group (SWG) meeting in February 2008 and at a full board meeting in June, the pair proposed ways to execute an end-run around the prohibition that Klein, as Prop 71's chief author, had promulgated when seeking the initiative's passage (Reynolds, 2008b). David Jensen of the *California Stem Cell Report* (CSCR) blogged that CIRM was holding what he termed a "mystery meeting" in February: "We can't tell you anything about it," CSCR reported:

> All we know is that the meeting has been scheduled. The agenda contains nothing more than the date and location. A few days ago, we asked CIRM about the nature of the meeting. So far, we have received no response.
>
> (CSCR February 17, 2008)

Susan Fogel of the ProChoice Alliance for Responsible Research (PCARR) reported similarly that,

> None of the meeting documents were posted on the CIRM website until I asked for them, and most were finally posted less than 24 hours before the meeting. Still, there was nothing to suggest that anything out of the ordinary was going to be discussed.
>
> (Fogel, 2008)

Fogel attended the meeting. A proposal made by Alan Trounson seemed to surprise almost everybody in the room: why not offer women undergoing fertility treatment a discount if they give up a number of their eggs to CIRM supported research? When Robert Klein produced a legal opinion supporting the proposal, arguing that contributing money to women's IVF treatment did not constitute compensation, only reimbursement, Fogel drew a reasonable conclusion. Klein and Tounson, she opined, had been "conspiring on turning California law on its head." Fogel surmised that Klein had secured the legal opinion from a law firm, "paid hundreds of thousands of dollars per year to provide legal advice to CIRM." For its part, CIRM suggested that the opinion was informal and spur of the moment (though the high-paid lawyer was conveniently present at the meeting).

Before ending the meeting, participants settled on five future agenda items. Four were about paying women for their eggs (Fogel, 2008).

The February 2008 meeting made very clear that securing access to eggs, despite the lack of a donor health registry, was a hurdle in need of jumping. CIRM President, Alan Trounson gave a rough estimate of how many eggs were needed to create just one stem cell line from nuclear transfer:

> At the present time for making an embryonic stem cell line from nuclear transfer in animals, including monkeys, is that it's going to take something like 100, 150, perhaps even 200 eggs in each case to make an embryonic stem cell line. It is certainly not arguably less than a hundred at the present time. The animal experiments would indicate that you are probably going to need around 100 eggs to make an embryonic stem cell line . . . .. Accessing those number of eggs is no trivial matter, no matter what the opportunities are.
>
> (CIRM, February 28, 2008, p. 108)

Would CIRM lean toward funding research involving nuclear transfer (egg intensive) or other forms of research requiring fewer eggs? Testimony at the meeting made clear that there was continuing demand for nuclear transfer. But asking women undergoing IVF to share some of their eggs presented an ethical problem because, as one presenter characterized: "[t]here would be no responsible IVF doctor who wouldn't take every single egg from that patient and mix it with sperm in the attempt to make fertilized embryos from those" (CIRM February 28, 2008, p. 125). Paying women directly for their eggs circumvented this conflict of interest but for this to happen, as SWG members understood, "we would have to go to the legislature" (p. 153).

What was CIRM policy? Even some of its members found CIRM policy in need of clarity. One member forthrightly shared her perplexity:

> Well, now I'm a little confused . . . .. So I want to say it back . . . to make sure I understand it. You want clarification, which we thought we had, that if a patient is undergoing reproductive therapy and decides to donate some of her eggs, we think as a group that she has the right to do so. If that's not clear in our yellow sheet, we want to make that clear. What we also agreed on in our group, I just want to be sure, is that she could not be paid for this.
>
> (p. 113)

Later, the member concluded that, "a lot of us were (confused)" (p. 153). And by the end of the meeting still felt that, "in the area of compensation, I still think that there is confusion as to what we agreed to do" (p. 234). Even if there had been clarity on what CIRM policy was, how was that policy being enforced? Was it being enforced at all? The committee seemed not to have a handle on that either, one of its members noting, "We have people who have submitted applications knowing the rules under which we're operating. And we left a lot of these responsibilities to the institutions to figure out how to implement these rules" (p. 115). Before the meeting adjourned, Fogel chastised the CIRM committee: "It's really distressing that nothing indicating this kind of discussion was on the agenda. There were no materials on the website until midday yesterday. And there are a lot of people who would be very interested in this" (CIRM, February 28, 2008, p. 149).

In 2010, an event dubbed "Incident at the Marriott" by the *California Stem Cell Report*, suggested how over the intervening two years, CIRM's desire for larger numbers of women's eggs had only simmered not evaporated – and that transparency about the issue had not improved. On June 14, Diane Beeson and Tina Stevens headed to San Francisco expecting to attend a CIRM workshop they heard about through the grapevine, "Procurement of Human Oocytes: What has been the Experience to Date?" Although the topic was of manifest social concern, CIRM offered no public announcement of the meeting. When Beeson and Stevens arrived at the venue, they were stopped at the door. The time of the meeting had been changed, a staffer told them. The meeting had already taken place. Moreover, they could not attend such a meeting, the staffer explained, because conferees needed to protect their intellectual property rights. Beeson and Stevens were stunned: What intellectual property concerns could there be over oocyte procurement policy? Why was the agenda switched/moved forward? Why wasn't a workshop concerning egg donation posted on the CIRM website in the first place when the topic is known to be of serious concern to the public, especially to women's health advocates? A CIRM communications officer defended the event in contradictory ways. After admitting that, "[a]ll of CIRM's scientific workshops are by invitation only," he insisted that, "[t]he women . . . were not told to leave." As the *Stem Cell Report* viewed it, "[t]his latest incident appears to be another case that does not reflect well on CIRM." It noted that the, "state attorney general's guide to California's open meeting law says agencies covered by the act are barred from imposing 'ANY CONDITIONS' on attendance at a meeting" (CSCR, June 2010).

The 2010 incident added to evidence suggesting that adequate transparency of CIRM's discussion of possible policy directions could not be assumed (*Nature*, 2012). It came as little surprise when in 2013, CIRM announced its desire to shift its policies in a way that would result in weakened safeguards for women providing eggs. It proposed moving beyond compensating women for the expenses they incurred when undergoing risky egg harvesting. It wanted to use taxpayer funds to finance research where women had been paid outright for their eggs.

No longer egg "donors," a targeted class of financially vulnerable college age women were becoming commercial bio-vendors in a free market characterized by inadequate egg donor studies and asymmetric information. This made meaningful informed consent impossible. The backsliding move would erode one of the only protections set in place by Prop 71 (which prohibited paying women for their eggs) and threatened the expanded protections insured by SB 1260, the Reproductive Health and Research law (see p. 80). CIRM's Standards Working Group (SWG), the entity responsible for establishing ethical requirements, gave notice of a July 23 meeting to discuss the proposal, including public input. Center for Genetics and Society, ProChoice Alliance for Responsible Research, Alliance for Humane Biotechnology, and Our Bodies Ourselves presented a joint memo (Appendix B). Representatives from some of the groups spoke, as did UC Berkeley medical anthropologist and founder of Organs Watch, Nancy Scheper-Hughes. Other critical groups not present but represented in public comment included, Black Women's Health Imperative, Breast Cancer Action, Cancer Prevention and Treatment Funds, Council for Responsible Genetics, National Women's Health Network, and We Are Egg Donors (CIRM Memo, July 24, 2013, p. 77).

When, after public comment, CIRM deferred voting on the proposal in favor of further discussion, it looked as though the critics' strong opposition had scuttled the move (Darnovsky, 2013; CSCR July 24, 2013). But maneuvers underway in the California legislature suggested that strategies designed to gain greater access to women's eggs radiated beyond CIRM into broader policy making arenas. Five months earlier, California Assembly Member Susan Bonilla (D, Concord) had introduced AB 926 (Schubert, 2013). Sponsored by ASRM, the "Reproductive Health and Research" bill would enable, as CIRM was seeking, paying women for their eggs. Critics, once again hobbled by lack of financing and dedicated time, mobilized to push back against the well-funded, lobbyist-backed stratagem to overturn laws protecting women's health. The Center for Genetics and Society sent its then Program Director, Diane Tober, to offer testimony at the Senate Health Committee. The Alliance for Humane Biotechnology arranged for Dr. Jennifer Schneider and Dr. Sindy Wei to make presentations. Dr. Schneider's daughter Jessica Wang, after donating eggs three times, died at the age of 31 from non-familial colon cancer (Appendix D). Dr. Sindy Wei nearly bled to death during egg donation when a prominent infertility specialist nicked one of her arteries while retrieving over 60 eggs from her ovaries (a woman usually produces only one or two eggs every cycle) (Appendix C; Tober, 2013). Senate Health Committee members ambling about the room chatting with each other during the authoritative, heart-wrenching presentations, gave clear indication of just how effective the bill's lobbying effort had been. AB 926 passed easily out of the Senate by a vote of six to one (CSCR, June 13, 2013). Critics organized to encourage a veto by the Governor, securing endorsement by 15 prochoice organizations (SynBioWatch, 2013). When the bill arrived at Governor Jerry Brown's desk in August, it met with his veto. "Not everything in life is for sale nor should it be," Brown declared. "The long-term risks are not adequately known. Putting thousands of dollars on the table only compounds the problem," he explained (Lifsher, 2013).

A disappointed Assemblywoman Bonilla later commented that Brown's conclusion was, "a very negative characterization and inaccurate." As she viewed it, she had just wanted to make it easier for scientists to get eggs for fertility research (Lifsher, 2013). But, in fact, it was never clear that the eggs that AB 926 would make more accessible would be used solely for fertility research. Despite the bill's title, "reproductive health and research," the type of research was not identified. The infertility industry's objective is to assist women suffering from infertility to conceive. Was an expanded market for women's eggs supposed to involve research on how to help infertile women conceive? If not that, if not chiefly that, then what?

For activists, a crucial question was how legislation like AB 926 would affect CIRM policy on acquiring eggs. Nothing about AB 926 made this clear. And CIRM policies, as we have seen, were difficult to decipher. On the one hand, the invention of iPS cells (described in Chapter 6) was seen as putting the demand for women's eggs in low gear (Thompson, 2013, p. 51). On the other hand, CIRM's 2013 move to pay women for their eggs suggested that CRIM funded research was still creating a significant demand for eggs. Given the porous relationship between infertility clinics and research cloning, would passing a bill such as AB 926 make it possible to fund researchers who gain access to eggs that they were prohibited from paying for directly? Would it make it easier for CIRM to return to the legislature seeking exemptions from their prohibitions on paying for eggs? There was little way of assuring that regulations protecting women's health were not being compromised. And still, no provisions were being set in place for long-term follow-up of donor health.

Despite the industry's 2013 failed attempt to expand the egg market through passage of AB 926, boosters wasted no time trying again. In 2016, Assemblywoman Autumn Burke (D-Inglewood) introduced AB 2531 (Senate Committee on Health, 2016). Once again, the infertility industry sponsored the bill, this one essentially the same as the one Governor Brown vetoed just three years before (CSCR, 2016). Testifying in opposition to the bill was the young co-founder of the advocacy organization, We Are Egg Donors, Raquel Cool. Her presentation drew a sharp contrast between the experience of many egg donors on the one hand, and infertility industry brochure statistics, on the other:

The industry claims OHSS is a preventable condition, but in 3 years I have seen only one case in which a doctor cancelled a cycle because of concern about protecting the donor. We are told that 10–20 eggs per cycle is the goal, and doctors know that greater numbers impose greater risks. But we routinely see numbers in the 30s, 40s, 50s and in some cases even up to the 70s or 80s. I can tell you, based on the experience of our nearly 1000 members, that moderate to severe OHSS occurs far more often than the 1% statistic that is being claimed.

(Appendix E)

The bill, which passed easily in the House and the Senate, never made it to the Governor's desk, expiring as "unfinished business" by the term's end (AB 2531). But oppositional critics were less than elated over the bill's mysterious demise. Earlier in the term, Diane Beeson called the Assemblywoman's office hoping to learn what she could about the bill's history and to make sure the office understood prochoice opposition arguments. Why would you introduce a bill so much like the one vetoed just three years before, she asked the staffer. The reply disclosed a juggernaut's stratagem: a veto presents no barrier to re-introducing the bill yet again. For well-moneyed interests with a lobbying arm there are no lost efforts, only continuing opportunities to "educate" legislators towards an eventual win.

In the meantime, continuing efforts to increase the market in eggs led to egregious transgressions of previously uncrossed ethical borders. In September 2017 a faculty member at San Francisco State University contacted activists to report that the University of California, San Francisco's Center for Reproductive Health had asked her to contact students via her listserv to help recruit egg donors for their program. The solicitation characterized this as a way for students to help pay the costs of tuition and other expenses. They could "donate" up to six times for $10,500 each time (Appendix F). The faculty member was thus encouraged to hold out to her students the promise of as much as $63,000, with no mention of health risks. For Our Bodies, Ourselves co-founder Judy Norsigian, this represented a new low:

[I]t solicits professors to use their authority and influence (and email lists of students) to encourage young female students to bolster their finances with a decision that could adversely affect their health. It is a deeply troubling request for faculty members to abuse their positions of power and trust.

(Norsigian and Richwald, 2018)

Coda

At the 2008 SWG meeting, CIRM heard medical testimony on the processes and risks of egg harvesting. It covered what was known about long-term risks of the drugs given to women undergoing the process:

> We know that infertile women, when they take these drugs over a long period of time, are not at risk for cancer . . . And . . . we feel quite safe in our knowledge that we can tell *infertile* [italics added] women that they don't have anything to worry about long-term. However, we don't have the same confidence simply because we don't have the same numbers in *fertile* [italics added] women. The stem cell research and egg donor research as it [sic] advances may allow us to have a larger cohort of women and may allow a registry that provides real numbers and that can lead to very viable conclusions.
>
> (CIRM, 2008, p. 169)

In other words, they didn't know much about the long-term health risks to young donors (fertile women) and they hoped to find out. A decade later, there is still no rigorously controlled tracking of long-term health risks to donors.

By 2016, bioresearch interests eager to use **CRISPR-Cas9** (see Chapter 6), touted as a way to "edit" genes, including those of eggs and sperm, and human embryos themselves, brought the possibility to CIRM's door. In February, the Institute's Standards Working Group considered whether CIRM should fund such research. It was a hot button issue. In April the year before, the National Institutes of Health announced that it would "not fund any use of gene-editing technologies in human embryos." Its Director, Francis Collins, explained, "the concept of altering the human germline in embryos for clinical purposes has been debated over many years from many different perspectives, and has been viewed almost universally as a line that should not be crossed" (National Institutes of Health, 2015). For CIRM, however, the question remained open (Darnovsky, 2016), and by early 2017 a committee empaneled by the U.S. National Academies of Sciences and of Medicine recommended research leading to eventual embryo modification, with the caveats that it be performed (according to a report in the journal *Science*), "only in rare circumstances and with safeguards in place."

The question's very consideration by CIRM raises unaddressed procedural issues: does the question of whether CIRM should fund projects that widen the door to human germline genetic engineering exceed its legislative mandate? Is such a highly consequential line of research properly left as a matter of an agency's administrative purview? When California citizens voted to create CIRM to find cures, did they mean to hand over human evolution and biosocial justice as well? Should visions of what a human future looks like be a matter of elite decision-making? Chapter 4 closes with these questions recognized but unanswered, serving as a gateway to a broader contextualization of the issues in Chapter 6. "The Road to Gattaca" lays out recent biotechnological developments and shifting cultural terrain that make the prospect of a society of "genetic haves and have nots" a matter of urgent consideration. This scenario may seem speculative, but is it not presaged by the current situation in which young women's bodies and health are being sacrificed in the pursuit of cures for others?

Before moving to Chapter 6, though, we take the next chapter, "Synthetic Biology: Extreme Genetic Engineering" to discuss an emerging toolkit of artificial genetic devices and the heated commercial efforts to apply them. Currently, synthetic biology is being used in plants, animals, and microbes. Additionally, via potential impacts on agriculture and fuel sources, it affects economies and societies as well. Beyond this, the ambitions of the players in this new field, many of whom come from the same world of embryo technology we have been describing, make these tools and methods central to proposals and prospective efforts to heritably modify humans.

Synthetic biology (sometimes called "synbio") technologies and their anticipated products, though not subject to the same controversies as those pertaining directly to the human body, raise potential problems of their own. To their funders, synthetic biologists trumpet the field's great promise. Cautions inadequately or never addressed include: threats to environmental safety, the potential rise of novel pathogens, and the monopolization of the food supply. For many critics watching the field, it only made sense to take seriously the stated desire to transform the human species and wonder whether counting the human costs of making the attempt would ever be acknowledged (Chapter 6).

Many of these issues converged on California's San Francisco Bay Area, the epicenter of the battle over Prop 71, and can be seen as a logical consequence of invigoration of the entrepreneurial victors of that fight. If vague promises about biotechnology could entice the public to turn over its tax revenues, similar strategies might be employed to appropriate municipal real estate.

Notes

1 Full text of required legal notice:

> Egg donation involves a screening process. Not all potential egg donors are selected. Not all selected egg donors receive the monetary amounts or compensation advertised. As with any medical procedure, there may be risks associated with human egg donation. Before an egg donor agrees to begin the egg donation process, and signs a legally binding contract, she is required to receive specific information on the known risks of egg donation. Consultation with your doctor prior to entering into a donor contract is advised.

2 Text of the Petition:

> Women undergoing egg extraction are not able to make informed choices about the risks involved, in part because long-term risks have not yet been adequately studied, especially for women providing eggs to be used by others. We urge the creation of a widely-publicized, privacy-ensured national registry to facilitate long-term tracking and long-term studies to better understand the risks of egg extraction, particularly with respect to the impact of drugs used for both suppression and stimulation of the ovaries. We also urge that advertisements and notices seeking women to supply eggs be required to state that long-term health risks of egg harvesting procedures are unknown.

Sources Consulted for Chapter 4

AB 926, "Reproductive Health and Research," California Legislative Information, California Legislature 2013–2014, regular session: http://leginfo.legislature.ca.gov/faces/billNavClient.xhtml?bill_id=201320140AB926

AB 1317, Assembly Bill CHAPTERED, 2009: www.leginfo.ca.gov/pub/09-10/bill/asm/ab_1301-1350/ab_1317_bill_20091011_chaptered.html

AB 2531, "Reproductive Health and Research," California Legislative Information, 2015–2016: http://leginfo.legislature.ca.gov/faces/billHistoryClient.xhtml?bill_id=201520160AB2531

"A Document from the Bioethics and Law Observatory Gives Recommendations on the Use of CRISPR Technique for Human Gene Editing": www.bioeticayderecho.ub.edu/en/document-bioethics-and-law-observatory-gives-recommendations-use-crispr-technique-human-gene-editing

Alliance for Humane Biotechnology, Center for Genetics and Society, Our Bodies Ourselves, and Pro-Choice Alliance for Responsible Research, "The Facts: A Bill to Permit Researchers to Pay for Eggs Is Bad for Women, Bad for California," May 30, 2013: www.geneticsandsociety.org/biopolitical-times/facts-bill-permit-researchers-pay-eggs-bad-women-bad-california

California's Stem Cell Agency Considers "Editing" Human Embryos: www.biopoliticaltimes.org/article.php?id=9173

California Stem Cell Report, "Day-long CIRM Meeting: No Topic Disclosed," February 17, 2008: http://californiastemcellreport.blogspot.com/2008/02/day-long-cirm-meeting-no-topic.html

California Stem Cell Report, December 7, 2008: http://californiastemcellreport.blogspot.com/2008/12/klein-says-he-cant-afford-to-work-for.html

California Stem Cell Report, "Incident at the Marriott: Stem Cell Agency Bars Public from Meeting," June 18, 2010: http://californiastemcellreport.blogspot.com/2010/06/stem-cell-agency-bars-public-from.html

California Stem Cell Report, April 10, 2012, quoting Marcy Darnovsky on CIRM conflicts of interest: http://californiastemcellreport.blogspot.com/2012/04/center-for-genetics-and-society-wrong.html

California Stem Cell Report, "Compensation for Human Eggs Approved by Key California Senate Committee, But Not for CIRM Researchers," June 13, 2013. http://californiastemcellreport.blogspot.com/2013/06/compensation-for-human-eggs-approved-by.html

California Stem Cell Report, "Limited Trounson Investigation Shows No Evidence of Illegal Conflicts," July 24, 2014: http://californiastemcellreport.blogspot.com/2014/07/limited-trounson-investigation-shows-no.html

California Stem Cell Report, "Pay-for-Eggs Legislation Up Again in California: Fertility Industry Trying to Repeal Ban on Compensation for Human Eggs in Research," April 18, 2016: http://californiastemcellreport.blogspot.mx/2016/04/pay-for-eggs-legislation-up-again-in.html

California Stem Cell Report, "'Less Than a Drop in the Bucket' – Dueling Perspectives on California's First Stem Cell Royalty Check," February 14, 2018: http://californiastemcellreport.blogspot.com/2018/02/less-than-drop-in-bucket-dueling.html

Center for Genetics and Society, "Open Letter to U.S. Senate on Human Cloning," March 19, 2002: www.geneticsandsociety.org/article.php?id=1958

Center for Genetics and Society, "Potential Conflicts of Interest at the CIRM," April 6, 2005: www.geneticsandsociety.org/article.php?id=318

Center for Genetics and Society, "SB 1260 (Standards for Egg Retrieval for Stem Cell Research) Fact Sheet," September 14, 2006: www.geneticsandsociety.org/article/sb-1260-standards-egg-retrieval-stem-cell-research-fact-sheet

Center for Genetics and Society, "Report of First Gene-Edited Human Embryos in the US," July 27, 2017: www.geneticsandsociety.org/press-statement/report-first-gene-edited-human-embryos-us

CIRM, "Before the Scientific and Medical Accountability Standards Working Group to the California Institute for Regenerative Medicine Organized Pursuant to the California Stem Cell Research and Cures Act, February 28, 2008: www.cirm.ca.gov/sites/default/files/files/agenda/transcripts/02-28-08.pdf

CIRM, "Before the Scientific and Medical Accountability Standards Working Group of the Independent Citizens' Oversight Committee to the California Institute For Regenerative Medicine Organized Pursuant to the California Stem Cell Research and Cures Act," Friday, July 25, 2008: www.cirm.ca.gov/sites/default/files/files/agenda/transcripts/07-25-08.pdf

CIRM, "Before the Scientific and Medical Accountability Standards Working Group to the California Institute For Regenerative Medicine Organized Pursuant to the California Stem Cell Research and Cures Act," December 12, 2008: www.cirm.ca.gov/sites/default/files/files/agenda/transcripts/12-12-08.pdf

CIRM, "Memo to CIRM Medical and Ethical Standards Working Group RE Consideration of Exception for Covered Stem Cell Lines," July 1, 2013: www.cirm.ca.gov/sites/default/files/files/agenda/130701_SWG%20Briefing%20Memo%20%2800200980-2%29.pdf

CIRM, "Before the Scientific and Medical Accountability Standards Working Group of the Independent Citizens' Oversight Committee to the California Institute For Regenerative Medicine Organized Pursuant to the California Stem Cell Research and Cures Act," July 24, 2013: www.cirm.ca.gov/sites/default/files/files/agenda/transcripts/1302407_SWG_TRANSCRIPT.pdf

CIRM, Statement Regarding Former CIRM President's New Position on a Stem Cell Company's Board, July 9, 2014: www.cirm.ca.gov/about-cirm/newsroom/press-releases/07092014/statement-regarding-former-cirm-presidents-new-position

CIRM, 2016 Annual Report: www.cirm.ca.gov/about-cirm/2016-annual-report

CIRM, "Before the Scientific and Medical Accountability Standards Working Group of the Independent Citizens' Oversight Committee to the California Institute For Regenerative Medicine Organized Pursuant to the California Stem Cell Research and Cures Act," February 4, 2016: www.cirm.ca.gov/sites/default/files/files/agenda/transcripts/StdsWkg Group-2-4-16%20Transcript.pdf

"Critics of California stem-cell agency address Institute of Medicine panel," April 13, 2012, *Nature* News Blog: http://blogs.nature.com/news/2012/04/critics-of-california-stem-cell-agency-address-institute-of-medicine-panel.html

Darnovsky, Marcy, "Unsupported Conclusions on Egg Procurement," *Biopolitical Times*, February, 13, 2007: www.geneticsandsociety.org/biopolitical-times/unsupported-conclusions-egg-procurement

Darnovsky, Marcy, "California Enacts Law to Reduce Risks to Women Who Provide Eggs for Stem Cell Research," March 6, 2007: www.geneticsandsociety.org/press-statement/california-enacts-law-reduce-risks-women-who-provide-eggs-stem-cell-research

Darnovsky, Marcy, "California Stem Cell Agency Delays Decision on Expanding the Market in Women's Eggs," *Biopolitical Times*, Center for Genetics and Society, July 25, 2013: www.geneticsandsociety.org/biopolitical-times/california-stem-cell-agency-delays-decision-expanding-market-womens-eggs

Darnovsky, Marcy, "California Stem Cell Agency Considers 'Editing' Human Embryos," *Biopolitical Times*, Center for Genetics and Society, February 9, 2016: www.geneticsandsociety.org/biopolitical-times/californias-stem-cell-agency-considers-editing-human-embryos

Davis, Rebecca, "China 'Clone Factory' Scientist Eyes Human Replication," December 1, 2015: http://news.yahoo.com/china-clone-factory-scientist-eyes-human-replication-061141389. html;_ylt=AwrXgiJen11WCkIAvoLQtDMD

Dutton, Diana B., *Worse than the Disease: Pitfalls of Medical Progress*, Cambridge, UK: Cambridge University Press, 1992.

Fikes, Bradley J., *The San Diego Union Tribune*, May 4, 2017: www.sandiegouniontribune. com/business/biotech/sd-me-cirm-president-20170503-story.html

Fogel, Susan. "Report from the Standards Working Group Meeting," *Biopolitical Times*, March 10, 2008: www.geneticsandsociety.org/biopolitical-times/report-cirm-standards-working-group-meeting

Friedberg, Errol C., *A Biography of Paul Berg: The Recombinant DNA Controversy Revisited*, Hackensack, NJ: World Scientific Publishing, 2014.

Futterman, Laurie G., and Louis Lemberg, "Cardiac Repair with Autologous Bone Marrow Stem Cells," *Am J Critical Care* 13 (2004): 512–518: www.ncbi.nlm.nih.gov/pubmed/15568657.

Graze, P.R., J.R. Wells, W. Ho, R.P. Gale, and M.J. Cline, "Successful Engraftment of Cryopreserved Autologous Bone Marrow Stem Cells in Man," *Transplantation* 27 (1979): 142–145: www.ncbi.nlm.nih.gov/pubmed/380074

Hall, Carl T., "STEM CELLS: The $3 Billion Bet / One Man's Scientific Mission / Housing Developer Leads California's Research Effort," *SF Gate*, April 11, 2005: www.sfgate.com/news/article/STEM-CELLS-The-3-Billion-Bet-One-man-s-2642279.php#photo-2121970

Hayden, Erika Check, "Stem Cells: Hope on the Line," *Nature*, July 2, 2014: www.nature. com/news/stem-cells-hope-on-the-line-1.15499

Hayes, Richard, "Human Genetic Engineering," Casey Walker, ed., *Made Not Born: The Troubling World of Biotechnology*, San Francisco, CA: Sierra Club Books, 2000.

Hiltzick, Michael, "California Stem Cell Agency Needs to Study Itself," *LA Times*, December 7, 2011: http://articles.latimes.com/2011/dec/07/business/la-fi-hiltzik-20111207

Hiltzick, Michael, "Did the California Stem Cell Program Promise Miracle Cures?" *LA Times*, May 30, 2012: http://articles.latimes.com/2012/may/30/news/la-mo-stem-cell-20120530

Hiltzick, Michael, "Research Firm Reaps Stem Cell Funds Despite Panel's Advice," *LA Times*, October 17, 2012: http://articles.latimes.com/2012/oct/17/business/la-fi-hiltzik-20121017

Hiltzick, Michael, "Conflicts of Interest Pervasive on California Stem Cell Board," *LA Times*, July 18, 2014: www.latimes.com/business/hiltzik/la-fi-hiltzik-20140720-column.html

Hoover Commission, State of California, Little Hoover Commission, "Commission Calls on State to Strengthen Stem Cell Board," June 26, 2009: www.lhc.ca.gov/studies/198/cirm/PressRelease198.pdf

Institute of Medicine, "Assessing the Medical Risks of Human Oocyte Donation for Stem Cell Research: Workshop Report," The National Academies Press, 2007: www.nap.edu/read/11832/chapter/1

Ivry, Dan, "Cell Out: Management Issues Plague Distribution of $3 Billion in State Stem Cell Research Funds," *LA City Beat*. March 3, 2005: www.lacitybeat.com/article.php?id=1837&IssueNum=94

Jensen, David, "Stem Cell Company Paid $443,500 to Former Head of State Agency that Funds Research," *The Sacramento Bee*, September 1, 2016: www.sacbee.com/news/state/california/article99432362

Jensen, David, "California's Stem Cell Agency Will Run Out of Money in Three Years. Should Voters OK Spending More?" *The Sacramento Bee*, January 17, 2017: www.sacbee. com/news/local/health-and-medicine/article126899909.html

Jewett, Christina, "Women Fear Drug They Used to Halt Puberty Led to Health Problems," *California Healthline*, February 2, 2017: http://californiahealthline.org/news/women-fear-drug-they-used-to-halt-puberty-led-to-health-problems/

Kaiser, Jocelyn, "US Panel Gives Yellow Light to Human Embryo Editing," *Science*, February 14, 2017: www.sciencemag.org/news/2017/02/us-panel-gives-yellow-light-human-embryo-editing

Knoepfler, Paul, "Review of Mitalipov Paper CRISPR'ing Human Embryos: Transformative Work on the Edge," *The Niche: Knoepfler Lab Stem Cell Blog*, August 2, 2017: https://

ipscell.com/2017/08/review-mitalipov-paper-crispring-human-embryos-transformative-work-edge/

Lee, Stephanie M., "Stem Cell Researchers Under Pressure to Produce," *SF Gate*, July 4, 2014: http://m.sfgate.com/technology/article/Stem-cell-researchers-under-pressure-to-produce-5600605.php

Leuty, Ron, "'Trounson affair' Another Strike Against California Stem Cell Agency," *San Francisco Business Times*, July 14, 2014: www.bizjournals.com/sanfrancisco/blog/biotech/2014/07/cirm-stem-cells-prop-71-alan-trounson-stemcells.html?page=all

Levine, Daniel S., "Strange Bedfellows Opposing Stem Cell Measure," October 7, 2004: www.bizjournals.com/sanfrancisco/stories/2004/10/11/story8.html

Levine, Judith, "What Human Genetic Modification Means for Women," World Watch, July/August 2002, pp. 26–29; and in Jeremy Gruber, and Sheldon Krimsky, *Biotechnology in Our Lives: What Modern Genetics Can Tell You About Assisted Reproduction, Human Behavior, Personalized Medicine, and Much More*, New York: Skyhorse Publishing, June 2013.

Lifsher, Marc, "Egg Donor Payment Bill Vetoed by Brown," *LA Times*, August 18, 2013: http://articles.latimes.com/2013/aug/18/business/la-fi-capitol-business-beat-20130819

Lomax, Geoffrey P., Zach W. Hall, and Bernard Lo, "Responsible Oversight of Human Stem Cell Research: The California Institute for Regenerative Medicine's Medical and Ethical Standards," *PLOS Med* 4(5) (2007, May): e114: www.ncbi.nlm.nih.gov/pmc/articles/PMC1858709/

McLean, Margaret R., "Red Light, Green Light: The California Cloning and Stem Cell Laws," *Markkula Center for Applied Ethics*: www.scu.edu/ethics/publications/ethicalperspectives/stemcelllaws.html

Mecoy, Laura, "Stem Cell Allies Divided Over Egg Collection," *The Sacramento Bee*, March 27, 2005, archived by Center for Genetics and Society: www.geneticsandsociety.org/article.php?id=1615

Memo, "To: CIRM Standards Working Group, From: Center for Genetics and Society, Pro-Choice Alliance for Responsible Research, Alliance for Humane Biotechnology, Our Bodies, Ourselves, July 24, 2013, Re: SWG Should Reject the Proposal to Allow Use of Cell Lines Created with Paid-For Eggs": www.scribd.com/document/155765833/CGS-July-2013-Letter-Opposing-Changes-in-CIRM-Cell-Line-Egg-Regulations

Memorandum of Points and Authorities in Support of Petition for Writ of Mandate and Alternative Writ of Mandate/Order to Show Cause (Undated, no case number), Paul Berg, PhD; Robert Klein; and Larry Goldstein, Petitioners vs. Kelvin Shelly, Secretary of State of California, Respondent, Geoff Brandt, State Printer; Bill Lockyer. Attorney General of California; Tom McClintock; H. Rex Green; John M.W. Moorlach; Judy Norsigian; Francine Coeytaux; Does I through X, inclusive, Real Parties In Interest. (Eventually, Case No. 04CS01015, Hearing Date, August 4, 2004.)

Munro, Neil, "Doctor Who? Scientists Are Treated as Objective Arbiters in the Cloning Debate. But Most Have Serious Skin in the Game," *The Washington Monthly*, November, 2002: www.washingtonmonthly.com/features/2001/0211.munro.html

National Institutes of Health, "Statement on NIH Funding of Research Using Gene-Editing Technologies in Human Embryos," April 28, 2015: www.nih.gov/about-nih/who-we-are/nih-director/statements/statement-nih-funding-research-using-gene-editing-technologies-human-embryos

National Library of Medicine, "Profiles in Science, The Paul Berg Papers: Molecular Biology and a Changing Academic Landscape, 1980–Present," https://profiles.nlm.nih.gov/ps/retrieve/Narrative/CD/p-nid/261

Nature News Blog, "Critics of California Stem-Cell Agency Address Institute of Medicine Panel," http://blogs.nature.com/news/2012/04/critics-of-california-stem-cell-agency-address-institute-of-medicine-panel.html

Norsigian, Judy, "Stem Cell Research and Embryo Cloning: Involving Laypersons in the Public Debates," *New England Law Review* 39 (2005): 527.

Norsigian, Judy, and Gary Richwald, "Risks vs. Reward: Two Women's Healthcare Professionals Offer Insight to the Dangers of Egg Harvesting," *Golden Gate Express*,

August 29, 2018: http://goldengatexpress.org/2018/08/29/should-professors-suggest-students-sell-their-eggs/

Official Voter Information Guide, California General Election, Text of Proposed Laws, Proposition 71: http://vig.cdn.sos.ca.gov/2004/general/propositions/prop71text.pdf; http://vig.cdn.sos.ca.gov/2004/general/english.pdf

Orange County Register, "Time to Target California Cronyism," July 21, 2014: www.ocregister.com/2014/07/21/editorial-time-to-target-california-cronyism/

Parisian, Suzanne, Open Letter, February 2005, online at: www.geneticsandsociety.org/article.php?id=181

"Petition for Human Egg Extraction Health Registry & for Warnings on Ads and Notices Seeking Egg Donors," 2011, AHB, OBOS, et al.: www.humanebiotech.org/sign-egg-donor-petition; and: www.ourbodiesourselves.org/take-action/egg-donation-health-risks-safety-data/

Piller, Charles, "California Voters Were Promised Cures. But the State Stem Cell Agency Has Funded Just a Trickle of Clinical Trials," STAT, January 19, 2017: www.statnews.com/2017/01/19/california-stem-cell-agency-cirm/

Pollack, Andrew, "Geron Is Shutting Down Its Stem Cell Clinical Trial," New York Times, November 14, 2011: www.nytimes.com/2011/11/15/business/geron-is-shutting-down-its-stem-cell-clinical-trial.html

Reuters Staff, "StemCells to Wind Down Operations After Ending Mid-Stage Study," May 31, 2016: www.reuters.com/article/us-stemcells-study-idUSKCN0YM1I2

Reynolds, Jesse, Submitted written testimony of Jesse Reynolds Director, Project on Biotechnology in the Public Interest, Center for Genetics and Society For the "Little Hoover" Commission on California State Government Organization and Economy, Sacramento, California, November 20, 2008a: https://lhc.ca.gov/sites/lhc.ca.gov/files/Reports/198/WrittenTestimony/ReynoldsNov2008.pdf

Reynolds, Jesse, blog post: "CIRM Won't Give Up on Eggs for Cloning Based Research," December 3, 2008b: www.geneticsandsociety.org/biopolitical-times/cirm-wont-give-eggs-cloning-based-work

Rodota, Joseph, and Bernard Munos, "To Fulfill Stem Cell Agency's Promise, Consider Winding It Down," The Sacramento Bee, February 2, 2017: www.sacbee.com/opinion/op-ed/soapbox/article130563309.html

RT, "Is Human Cloning Here? Chinese Scientist Ready to Clone People at His 'Replication Factory'," December 1, 2015: www.rt.com/news/324110-china-human-cloning-factory/

San Francisco Examiner, "California Stem Cell Czar Offered Top Pay, Asked to Name Job Description," May 29, 2011: www.sfexaminer.com/california-stem-cell-czar-offered-top-pay-asked-to-name-job-description/

SB 1064: California Stem Cell Research and Cures Act, 2009–2010, Senate Bill No. 1064: http://leginfo.legislature.ca.gov/faces/billNavClient.xhtml?bill_id=200920100SB1064

Schubert, Charlotte, "California Bill Poised to Raise Restrictions on Egg Donation," Nature, June 18, 2013: www.nature.com/news/california-bill-poised-to-lift-restrictions-on-egg-donation-1.13218

Senate Committee on Health, Senator Ed Hernandez, O.D., Chair, Bill No. AB2531, Author, Burke, Version February 19, 2016, Hearing Date, June 8, 2016, Consultant, Melanie Moreno.

Servick, Kelly, "Failed Spinal Cord Trial Offers Cautionary Tale," Science 355(6326) (February 17, 2017): https://d2ufo47lrtsv5s.cloudfront.net/content/355/6326/679/tab-article-info

Shanks, Pete, "Shameful Conflicts of Interest Involving California Stem Cell Agency," July 9, 2014: www.geneticsandsociety.org/biopolitical-times/shameful-conflicts-interest-involving-californias-stem-cell-agency?id=7891

Simpson, John M., "Consumer Watchdog Welcomes Institute of Medicine Report Calling for Sweeping Reforms at California Stem Cell Agency," Press Release for Consumer Watchdog, 2012: www.consumerwatchdog.org/newsrelease/consumer-watchdog-welcomes-institute-medicine-report-calling-sweeping-reforms-california

Stein, Rob, "Exclusive: Inside the Lab Where Scientists Are Editing DNA in Human Embryos," August 18, 2017, NPR, KQED Public Media, www.npr.org/sections/health-shots/2017/08/18/543769759/a-first-look-inside-the-lab-where-scientists-are-editing-dna-in-human-embryos

SynBioWatch, "Coalition Urges VETO of AB 926 – Human Egg Harvesting Threatens Women's Health": www.synbiowatch.org/2013/06/coalition-urges-veto-of-ab-926-human-egg-harvesting-threatens-womens-health/?lores

Thompson, Charis, *Good Science: Ethical Choreography of Stem Cell Research*, Cambridge, MA: MIT Press, December 20, 2013.

Tober, Diane. "Update: California Bill Would Overturn Protections for Women Providing Eggs for Research," *Biopolitical Times*, June 13, 2013: www.geneticsandsociety.org/biopolitical-times/update-california-bill-would-overturn-protections-women-providing-eggs-research?id=6947

Vaughan, Christopher, and Kevin Cool, "Cell Division," *Stanford Alumni*, May/June 2003: https://alumni.stanford.edu/get/page/magazine/article/?article_id=36663

"Voter Information Guide," California General Election, November 2, 2004, Proposition 71 text, pp. 147–155; Analysis by Legislative Analyst, pp. 68–71; Argument in Favor of Proposition 71, Rebuttal to Argument of Proposition 71, Argument Against Proposition 71, Rebuttal to Argument Against Proposition 71, pp. 72–73: http://vig.cdn.sos.ca.gov/2004/general/english.pdf

Wadhwa, Vivek, "If You Could 'Design' Your Own Child, Would You?" *The Washington Post*, July 27, 2017: www.washingtonpost.com/news/innovations/wp/2017/07/27/human-editing-has-just-become-possible-are-we-ready-for-the-consequences/?utm_term=.92117a512461

We Are Egg Donors (WAED): www.weareeggdonors.com/blog/ohss

Witherspoon Council, "The Stem Cell Debates: Lessons for Science and Politics," *The New Atlantis: A Journal of Technology and Science* (Winter 2012): www.thenewatlantis.com/publications/the-stem-cell-debates-lessons-for-science-and-politics

Wobus, A.M., and K.R. Boheler, "Embryonic Stem Cells: Prospects for Developmental Biology and Cell Therapy," *Physiol Rev* 85 (2005): 635–678.

Worth, Katie, "California Stem Cell Czar Offered Top Pay, Asked to Name Job Description," *San Francisco Examiner*, May 29, 2011: www.sfexaminer.com/california-stem-cell-czar-offered-top-pay-asked-to-name-job-description/

Wright, Susan, *Molecular Politics: Developing American and British Regulatory Policy for Genetic Engineering, 1972–1982*, Chicago, IL: University of Chicago Press, 1994.

Yes on 71 archived website: http://digital.library.ucla.edu/websites/2004_996_027/index.htm; http://digital.library.ucla.edu/websites/2004_996_027/quotes.php.htm; http://digital.library.ucla.edu/websites/2004_996_027/tv_radio.php.htm; http://digital.library.ucla.edu/websites/2004_996_027/documents/KeyTalkingPoints.pdf; http://digital.library.ucla.edu/websites/2004_996_027/news_clip_0824_update.php.htm

5
Synthetic Biology
Extreme Genetic Engineering[1]

What if we could liberate ourselves from the tyranny of evolution by
being able to design our own offspring?

> (Drew Endy, bioentrepreneur, Assistant Professor
> of Bioengineering, Stanford University,
> Specter, 2009, p. 61)

[W]hy stop with microbes? It will soon be possible to make entirely
novel forms of plants or animals (including man).

> (Johnjoe McFadden, Professor of Molecular Genetics,
> University of Surrey, UK, McFadden, 2009)

An October 2004 news feature in the international science magazine, *Nature*
offered an arresting declaration: genetic engineering is old hat. A new field is
emerging, the accompanying editorial explained, and unlike what came before,
"synthetic biology," was "no longer a matter just of moving genes around."
Instead, it was, "shaping life like clay" (Ball, 2004; *Nature*, 2004). Old recombinant DNA techniques sought to modify existing organisms by inserting or
altering individual genes. Synthetic biologists, by contrast, were seeking not
merely to tweak an existing organism to enhance or harness its original function. They sought instead to create altogether novel life forms with wholly
unprecedented functions. Using parts of existing organisms as raw material in combination with synthetic chemicals, they sought to remodel life to
unique, patentable purpose. At the time of the *Nature* editorial, bacteria and
yeast already had been designed to build proteins not found in nature, possessing properties unlike anything seen previously on earth. Fashioned from parts
found in nature but not evolved by nature, these creatures were, "synthetic."

Just how far do imaginations reach? As synthetic biologists tend to
view it, all life can be deconstructed, reconstructed, and commodified.
The only limitations are those of human imagination and the vicissitudes
of moral sensibilities. Beyond new kinds of commodities and consumables, the ambitions of the field's visionaries tilt toward production of

"improved" humans. Drew Endy's dream to "liberate" the human species from evolution (see Chapter epigraph) exemplifies the agenda of deconstructing and reconstructing human biology. Similarly, Harvard's George Church related to a reporter that he, "wouldn't mind being virus-free." The interviewer explains that, while it "may be too late to reengineer all of his own cells to prevent viral infections . . . Church doesn't rule out the possibility of rewiring the genome of a human embryo to be virus-proof" (Bohannon, 2011). By 2012, bioentrepreneur Craig Venter had famously proclaimed, "Not too many things excite my imagination as trying to design organisms – even people" (Sharp, 2012).

Synthetic biology strategies are not currently being brought to bear on human reproduction.[2] But declarations of synthetic biology's leaders to do so make it more than a tool. It makes it a transformational force of all biological life including, potentially, human life. And it provokes a profound question: should bioentrepreneurial instincts be the ones taking us to the future? Finding the appropriate place of biotechnology in civic life makes it worth connecting dots between strategies driving synthetic biology, the social ramifications of those strategies, and the common aspects between the field of synthetic biology and other biotechnologies, especially reproductive biotechnologies (Chapter 6).

While some synthetic biology researchers draw lessons from evolution, their preference is to circumvent it. The goal is to engineer around ancient constraints to achieve instrumental objectives (Skerker et al., 2009; Nandagopal and Elowitz, 2011). In a real sense, synthetic biology treats cellular life as a programmable "black box" within which ingenious functionalities can be constructed. This is reminiscent of the iPhone and other Apple-manufactured mobile devices. Steve Jobs contrived them to be black boxes by making the hardware difficult to access and the software closed source, but simultaneously to be highly amenable to the development of "apps," i.e., narrowly focused applications (Isaacson, 2011).

"Extreme genetic engineering" and "genetic engineering on steroids" are monikers coined by industry watchers to characterize the field. The appellations highlight the way synthetic biology departs from "old hat" recombinant DNA techniques – and from the practice of doing science. Practitioners include computer scientists and engineers as well as bioscientists. Together with investors, they develop biotechnologies wed inextricably to mercantile agendas and, often, an industrial platform. Visions of commercial applications fuel imaginations, patent seeking, and new bio-industries. Synthetic biology rests on a mechanistic conceptualization of life. Living things, from this perspective, are entities possessed of interchangeable parts (ETC Group, 2007). Enabled by this tunnel-vision perspective, practitioners slice the sharp edges of engineering strategies through the complexities of microorganisms, hijacking their nature and assigning them new functions.

Sheltered under the ill-fitting notion of life as machine, synthetic biologists find cover for excising sub-cellular arrangements and swapping them out for artificially created elements whimsically termed "standard molecular parts" or "biobricks." In the first flush of the synthetic biology initiative the Registry of Standard Biological Parts (RSBP) associated with the non-profit BioBricks Foundation (now apparently both defunct) promised "a collection of genetic parts that can be mixed and matched to build synthetic biology devices and systems" (Newman, 2012). But tool bucket metaphors function better as tropes than they do in practice.

Current understanding of molecular genetics and cellular biochemistry guarantees that many such parts will not exhibit the same activity in different assemblages (see Chapter 2). This goes unmentioned in the online outreach materials the Foundation prepared for the public and business communities. According to a report in the journal *Science*, participants at a synthetic biology meeting in July 2010 concluded that of the 13,413 items then listed in the RSBP, 11,084 didn't work. One presenter noted, "Lots of parts are junk" (Kean, 2011). When figures of speech fall short of real world performance, it raises questions. What kinds of unintended consequences can result? Which individuals, social groups, and natural environments will be left to bear the risks? With such questions in mind, industry watchdogs published *The Principles for the Oversight of Synthetic Biology* in December 2012 (Friends of the Earth, 2012).

Organized by Friends of the Earth (FOE), the Erosion, Technology and Concentration Group (ETC), and the International Center for Technology Assessment (ICTA) *Principles* was endorsed by 117 civil society groups from around the globe. The sixteen-page document offered a set of governance principles and synthetic biology-specific regulations to protect the environment and ensure worker safety. At the center of the declaration is adherence to the Precautionary Principle:

> When an activity raises threats of harm to human health or the environment, precautionary measures should be taken even if some cause and effect relationships are not fully established scientifically. In this context the proponent of an activity, rather than the public, should bear the burden of proof. The process of applying the Precautionary Principle must be open, informed and democratic and must include potentially affected parties. It must also involve an examination of the full range of alternatives, including no action.

If developed subject to the precautionary principle and implemented in reversible ways with appropriate safety monitoring and controls, synthetic biology can potentially contribute in a modest fashion to science and medicine. Some of the apps devised by synthetic biologists are already useful as tools for genetic research (Constante et al., 2011), and others show potential as diagnostic sensors for abnormal blood proteins and environmental toxins, and as therapeutics against cancer (Huang et al., 2008; Haynes and Silver, 2011; Auslander et al., 2011). But, there continues to be a lack of proper oversight and risk assessment. With insufficient attention paid to risks, environmental safety, or social disruptions caused by shifting markets, proponents implement their synbio dreams: why not engineer algae to produce gasoline, design plants so that they glow, or tweak yeast to produce vanilla? In 2008, bioentrepreneur Jay Keasling shared what synthetic biology meant to him: "anything that can be made from a plant, can now be made by a microbe in a vat" (Thomas, 2008).

Synthetic biologists have a variety of ways to create new organisms (FOE, n.d.). Bio-products might be specified by introducing modified genes into familiar organisms, by unprecedented multi-gene molecular pathways added to these life-forms, or in some futuristic but not entirely far-fetched projections, by placing new genes and pathways in novel types of living systems. A prevailing synthetic biology method employs a computer program to spit out unique synthesized sequences of DNA's four nucleotides – cytosine, guanine, adenine, and thymine (C, G, A, and T). Using one of several genetic engineering techniques, these strands are then inserted into existing organisms. By such re-mixings, exotic self-replicating microorganisms can be conjured, harvested, and put in the service of reinvented production processes, fermenting in industrial vats. Consumables range from cleaning products, cosmetics, fragrances, flavorings, and pharmaceuticals, to fuels, plastics, and industrial chemicals. What do microorganisms fermenting in vats eat? As synthetic biology commercial interests move to create a "biomass-based economy," *Principles* explains, "any type of plant matter can be used as feedstock for tailored synthetic microbes to transform into high value commercial products – anything from fuels to plastics to industrial chemicals" (FOE, 2013). This may sound like a do-able prospect. But there are serious environmental and social consequences.

In the first place, there isn't enough land or plant matter for everything being envisaged. Increasingly large quantities of plant material will be necessary as the shift to biomass-derived feed stocks gets underway. This means that most of the needed plant matter will be extracted or cultivated in the global south. *Principles* drives home what this means: disruption of fragile ecosystems, exacerbation of damage done from industrial crop production, and further pressure on the short supply of land. And there are human costs. Some synthetic biology projects,

> propose to replace botanical production of natural plant-based commodities (e.g., rubber, plant oils, herbal medicines like the anti-malarial artemisinin) with vat-based production systems using

synthetic microbes or to move production to genetically engineered plants ... [T]hese substitutions could have devastating economic impacts on farming, fishing and forest communities who depend on natural compounds for their livelihoods.

(FOE, 2013)

Despite the inadequately considered social implications of these shifts, a wide array of products made using synthetic biology methods already lurk undetectably in the market place. More are headed to retail. None of these products bear labels identifying their biosynthetic origins. In 2009, the premiere international insurance company, Lloyd's of London, released an emerging risks report on synthetic biology. It urged the industry to establish an international system of tracking and labeling of synthetic biology products. "Labelling," it advised, "will give consumers choice and may reduce litigation costs if damages arise" (Lloyd's, 2009). But industry leaders differ sharply. In 2014, Friends of the Earth learned that their recommended PR strategy was to avoid using the terms "synthetic biology" or "genetically engineered" altogether. These identifiers, strategists argued, would engender public backlash. Instead, if descriptors are needed, the phrases, "fermentation derived" and "nature identical" would do, they proposed (FOE, 2014).

Although *Nature*'s 2004 editors cautioned practitioners to attend to safety and secure public understanding, they did not call for new regulations. Untroubled by the decades of genetically modified organisms that had, "no doubt escaped into the environment," old rules of the road would, presumably, serve well enough. After all, none of the escapees had produced, "a biological Chernobyl," they assured. Instead, wanting to secure public trust, they grasped at the appearance of concern. "What will happen if biologists announce that they have made the first living cells from scratch without having demonstrated to the public any concern for the implications?" The editors acknowledged that members of the synthetic biology community had begun considering the risks and ethical implications of their undertakings. But they wanted more. They wanted another Asilomar. (As recounted in Chapter 1, this was the 1975 meeting at which scientists agreed to perform gene splicing experiments under self-imposed guidelines.) The editorial was not, however, spotlighting a need for regulations. The Asilomar conference, they urged, "achieved something potentially more valuable: wide press coverage ... won the public's trust that scientists were behaving responsibly" (*Nature*, 2004). Only public trust could stave off anticipated calls for enforced restraint.

The editors needn't have worried. Just as CRISPR-Cas 9 gene editing of human embryos is underway without broad public consensus (Chapter 6), the public is scarcely aware of the assortment of synthetic biology products already in markets. Synthetically fermented alternatives to coconut oil and cocoa butter derivatives, stevia, and saffron are among the earliest products where synthetically created ingredients have slipped into foods and onto grocers' shelves (FOE, 2014). ETC Group is the leading civil society watchdog organization tracking developments in synthetic biology. It characterizes the field as a multi-billion-dollar industry with synthetic biology produced ingredients in soft drinks, soaps, face creams, and detergents. These elements are, they explain, "unregulated, unlabeled and under the radar of public awareness." Approximately "a hundred SynBio companies have deals with the largest chemical, food, energy and cosmetics companies on the planet," according to ETC, "many of them . . . household names" (ETC, video). Swift and arcane, the approval processes for synthetic biology methods and products might as well have been wholly clandestine. The Woodrow Wilson Center's Synthetic Biology Project acknowledges that,

> [a]fter more than twenty years of basic and applied research, applications based on synthetic biology are gaining in commercial use. But it has been difficult to find out how many synbio products are on the market or may potentially enter the marketplace in the near future.
>
> (Woodrow Wilson Center, 2018)

The difficulty of assessing synthetic biology's impact on the marketplace is but one of its uncontainable aspects. The field encourages an ethos of hacking and gamerism. This results from its implicit and sometimes explicit view of cells and multicellular organisms as devices that implement sets of defined tasks or functions, or if they are not inherently so, can be engineered to fit this description. A 2011 profile of the Harvard synthetic biology researcher George Church, for example, is titled "The Life Hacker" (Bohannon, 2011). Some of the incubators are the International Genetically Engineered Machine (iGEM) undergraduate competition, held yearly at the Massachusetts Institute of Technology (MIT), and venues of the do-it-yourself biology (DIY bio) movement (Kean, 2011). The genetically modified microbes and animal and plant cells generated by these activities, as well as those in ostensibly more accountable academic and industrial settings, could potentially invade and establish themselves in surrounding ecosystems in a disruptive and destabilizing fashion.

The DIY bio movement represents serious risk for both synthetic biology and gene editing. (There is no consensus on where these fields overlap. Where synthetic biology involves engineering of biochemical pathways and production of biological novelties, gene-editing technology (Chapters 2 and 6) involves simpler changes – base substitutions leading to one amino acid substitution, for example. But the line between synthetic biology and gene editing is not sharp, and experts argue about what constitutes a novelty.) DIY bio thrives in the crevices of an ineffectual regulatory patchwork characterized aptly in a May 2018 *New York Times* exposé on DIY gene editing:

Authorities in the United States have been hesitant to undertake actions that could squelch innovation or impinge on intellectual property. The laws that cover biotechnology have not been significantly updated in decades, forcing regulators to rely on outdated frameworks to govern new technologies. The cobbled-together regulatory system, with multiple agencies overseeing various types of research, has left gaps that will only widen as the technologies advance.

(Baumgaertner, 2018)

Even an 18-year-old DIY bio enthusiast had to agree that, "[t]hese regulations aren't going to work . . . when everybody has a DNA synthesizer on their smartphone." Mirroring professional science culture, biohackers "self-regulate." An FBI "biological countermeasures specialist" told the *Times* that they rely on biohackers themselves to report suspicious behavior. "But," said the Director of the Johns Hopkins Center for Health Security, "if you really want to do this, there isn't a whole lot stopping you."

There is immense interest in synthetic biology by major corporations in the energy, chemical, and agricultural sectors. Venture capitalists perpetually hover around the relevant laboratories determined to recruit publicly funded academic scientists into startup operations (Kunjapur, 2015). According to one industry analyst, synthetic biology startups in the biofuels and bio-based chemicals sector received $1.84 billion in private funds between 2004 and 2010. According to sources in the same report, four of the top five energy companies (Royal Dutch Shell, Exxon Mobil, British Petroleum, and Chevron), three of the top five chemical corporations (BASF, Dow, and Exxon Mobil), and the top three grain trading companies (Cargill, Archers Daniel Midland, and Bunge) had established partnerships with synthetic biology companies by 2011. By 2017, food, flavor, fragrance, textile, and cosmetic firms were leading the partnerships with over 410 synthetic biology companies. Trade group Synbiobeta reported an annual $1.8 billion cash injection by venture capital that same year, a market predicted to hit $38 billion in sales by 2020 (Synbiobeta, 2018; MarketWatch, 2018).

The U.S. military takes a keen interest in the technology. At a 2011 conference convened by the Department of Defense (DOD)'s Defense Threat Reduction Agency synthetic biologists were asked to look for "more environmentally friendly ways to manufacture explosives." One University of Texas researcher who holds grants from several DOD agencies asserted, "You can have someone die because of the way we currently prepare munitions, or you can put in proposals to try to make it easier and safer and greener to make munitions." Another bioengineer, from Boston University, was awarded $1.5 million a year for five years by the DOD's Office of Naval Research to develop "microbiorobots" bacteria genetically programmed to sense chemicals in the environment. Such robots could potentially be repurposed to deliver poisons or explosives (Hayden, 2011).

Craig Venter, the principal private entrepreneur behind the Human Genome Project of the 1990s, has entered into is a $600 million program sponsored by Exxon to engineer the genome of algae, purportedly to solve the global climate and food crises. However, as noted by the ETC Group, "Since [algae] are very common in the environment, there is a possibility of outcrossing with the natural species and contamination of microbial communities in soil, seas and animals, including humans," they explain. "Microbes propagate and mutate quickly and also move through soil, waterways and other routes, so it may be especially difficult to track escapes." They make plain that the unintended consequences of reengineering the biology of algae or changing algal stocks on any large scale globally could be catastrophic. Doing so may, "directly impact the global oxygen cycle, carbon cycle, nitrogen cycle and ozone production potentially in unpredictable and harmful ways." The report found no comfort in assurances offered up by the industry:

> Synthetic biologists contend that their lab-made creations are probably too weak to survive outside the optimized conditions in which they were developed; however, this assumption has been proven wrong before [for **transgenic** corn and soy].
>
> (ETC Group, 2010, 38, 49)

ExxonMobil bullishly promotes algal fuel even in advance of its assumed commercial feasibility. Television ads combine clever visuals and fairytale music with a voiceover narration of "the story" of algae delivered in rhyming meter. Evidently, it is an attempt to sell nothing to the viewer but normalization of the idea: "Once there was an organism so small, no one thought much of it at all. People said it was just a mess until ExxonMobil scientists put it to the test." The nonprofit Biofuelwatch tells a different story. "Scientists are clear that GE microalgae will inevitably escape from cultivation facilities," states Biofuelwatch Co-Director, Rachel Smolker. "Many of the traits that are being engineered to create algal 'chemical factories' could result in their outcompeting and proliferating out of control in the wild." Friends of the Earth's Senior Food and Technology Campaigner, Dana Perl concurred. "Rushing genetically engineered algae into production ahead of safety assessments and oversight could result in serious unintended consequences," she warned. "These organisms could become 'living pollution' that is impossible to recall" (Biofuelwatch, 2017).

The thriving synthetic biology hacker culture is only voluntarily regulated (which means, in effect, not at all) despite the potential and intention to produce increasingly exotic life forms (Kean, 2011). Bioentrepreneurs often express frustration with the regulatory inconveniences that were set in motion because the public and governmental agencies took seriously the concerns of an earlier generation of scientists (see Chapter 1). Many are determined to weaken or eliminate the existing rules and prevent further restrictions on emerging technologies in the field (Pollack, 2010). "Self regulation" is the watchword, and the bureaucratic tendency is to subsume the microbes and other organisms prospectively created by synthetic biology under the NIH guidelines, even though the technology addressed by those guidelines is very different from that currently available with synthetic biology. But while bioentrepreneurs favor self-regulation, industry watchers and watchdogs keen on safety find present guidelines far from satisfactory. For Jim Thomas of the ETC Group, believing that voluntary guidelines from the era of early genetic engineering are sufficient for synthetic biology is like assuming that rules for horse and buggy operation are adequate in the age of the automobile (personal communication to TS, 2011).

Similarly, although Lloyd's of London's 2009 evaluation acknowledged synthetic biology's potential, it also offered striking admonitions. "Risk assessments must consider the impact if all controls fail," it warned:

> If the consequences to the ecosystem are potentially severe then contingency plans should be put in place to limit damage. In the most extreme cases it must be questioned whether the risks are worth taking ... For organisms that are planned to be released into the environment this is particularly critical.
>
> (Lloyd's, 2009)

The report explicitly acknowledged the cautionary history of early genetic engineering: "there has been unexpected gene flow from both approved and experimental GM crops into conventional food, sometimes leading affected farmers to have to dispose of 'contaminated' crops at great expense." It also recalled lessons to be learned from the 1980s food crisis triggered by the unanticipated "mad cow disease" (bovine spongiform encephalopathy) (Lloyd's, 2009). In a particularly well-targeted critique, it noted that,

> we are far from understanding how genomes work ... The danger is that we take action based on our understanding now to find later that there were unintended and unimagined consequences. It is possible that two or more benign strands of DNA will interact so that the risk is far greater than the sum of the parts.
>
> (Lloyd's, 2009)

Ultimately, the authors advised insurers to "aim for a precautionary and consistent global approach," counseling them to "take part in a broad debate on the use of this new technology before it becomes embedded." But is broad debate over stealth applications even possible? The report's advice, seemingly in earnest, was the belated suggestion to lock the barn door after noticing the proverbial horse had galloped far afield.

The Politics of Synthetic Biology

The history of synthetic biology's acceptance in the United States is not a tale of broad debate. In May 2010, the Obama administration charged the Presidential Commission for the Study of Bioethical Issues to prepare a report on the ethical implications of synthetic biology. Although the Commission's work was in some sense public, its convenings were elite and inaccessible. By December of that year the report had already been submitted. With minimal discussion, it dispensed with important issues such as the possibility of novel, disruptive microorganisms escaping into the environment. The Commission acknowledged that, "creating new organisms that have uncertain or unpredictable functions, interactions, and properties could affect ecosystems and other species in unknown and adverse ways." It also recognized that, "[t]he associated risks of escape and contamination may be extremely difficult to assess in advance, as such novel entities may have neither an evolutionary nor an ecological history" (New Directions, 2010, p. 56). But, despite acknowledging concern over such dire risks, it accepted uncritically proponents' presumably reassuring assertion that "synthetic organisms allowed to develop in the laboratory have consistently evolved toward nonfunctionality" (p. 57; New Directions, 2010, p. 70). Nor did it address how even if escaped microbes are short-lived, demonstrated transfer of their novel genes and genetic networks to environmentally established counterparts was possible (Heuer and Smalla, 2007). The Commission's 13 scientists, policy experts, and ethicists endorsed the report unanimously, giving synthetic biologists a green light to self-regulate. New regulations would not be required (Pollack, 2010).

It was up to civil society groups to provide critical scrutiny and demands for adequate regulatory regimes for the more than 200 academic and commercial laboratories engaged in synthetic biology research in the U.S. Shortly after the release of the Presidential Commission's 2010 report, Friends of the Earth U.S. (FOE), the Erosion, Technology and Concentration Group (ETC), and the International Center for Technology Assessment (ICTA) drafted a letter of protest. Fifty-eight environmental, public interest, and religious groups from 22 countries signed the document (Appendix G). The letter cited the report's ignoring of the precautionary principle, lack of adequate concern for the environmental risks of synthetic biology, dependence on unsubstantiated technologies for environmental safety, and reliance on the mirage of self-regulation. "Self regulation," it charged, "means no real regulation or oversight of synthetic biology." The signatories called for a moratorium "on the release and commercial use of synthetic organisms until a thorough study of all the environmental and socio-economic impacts of this emerging technology has taken place." It should remain in effect, they demanded, "until extensive public participation and democratic deliberation have occurred on the use and oversight of this technology" (Appendix G). Authorities ignored the petition.

The following year held even more sobering news. "Lab Fight Raises Security Issues," headlined the *New York Times* in October 2011. Reportage quickly tied the horrors of 9/11 and the anthrax attacks following in its wake to mounting unease over synthetic biology. DNA sequences necessary for creating designer organisms were stored on public databases and the specter of maniacal individuals gaining access unsettled close observers. It cast a pall over the discipline, opined one Rutgers University chemical biologist interviewed. Where the article's headline sounded an alarm over biosecurity, its analysis tapped into an underlying issue pulsing at the heart of synthetic biology research: the inadequacies of self-regulation. Tensions between a coalition of biolabs administered on the UC Berkeley campus and a UC Berkeley professor hired to evaluate the program, for example, were bringing the concern into sharper focus.

The Synthetic Biology Engineering Research Center (Synberc) was a self-described major U.S. research program with a goal to "make biology easier to engineer." A 10-year National Science Foundation (NSF) grant in the neighborhood of $23.3 million dollars launched the enterprise in 2006. (The Engineering Biology Research Consortium (EBERC) superseded it in 2016.) The award came with the requirement that the coalition study social aspects of its work, including ethics and security. They hired UC Berkeley anthropologist and biosafety expert Paul Rabinow to undertake that effort. Within four years, that relationship soured. In March 2010, an NSF evaluation found that Synberc's attention to security and risk was inadequate and controversy ensued over who was to blame. Synberc's strategic director, bioentrepreneur and assistant professor of bioengineering at Stanford, Drew Endy, replaced Rabinow on the project. Rabinow continued his research at Synberc until July 2010 at which time he resigned from the center altogether (Gollan, 2011). Conflicting characterizations of what had gone wrong highlight the uncharted ramifications of what it means to allow bioentrepreneurs to self-regulate (Rabinow and Bennett, 2012).

Synberc's director Jay Keasling, was also chief executive of the U.S. Department of Energy's Joint BioEnergy Institute (JBEI) and a professor in the chemical engineering and bioengineering departments at UC Berkeley. A bioentrepreneur, Keasling also launched a number of companies to commercially develop applications. He was a co-founder of Condon Devices, Amyris, and LS9. In February 2012, he would found Lygos, Inc. to commercialize JBEI research on biofuels (Jones, 2012). Keasling laid full blame at Rabinow's doorstep. Rabinow, as Keasling viewed it, did not adequately communicate his ideas or implement them. In contrast Rabinow, who demurred that his work was not difficult to comprehend, confided that he resigned after becoming fed up with Synberc's indifference to their "responsibility to larger society, which is funding them, by entrusting them to manipulate life." He shared that "how profoundly irresponsible these guys are," began to worry him. "There are possibilities of all kinds of nefarious things happening. There is no reason that someone couldn't modify a virus; you could release it on an airplane or subway, and it could have profound terror effects," he stated. "DNA synthesis companies have no way of currently telling, once the sequences are put together, what the result will be. Somebody could manufacture pathogens that are dangerous to the environment," he explained. Synberc maintained that their research used the same DNA and proteins used since the 1970s. No new biological or chemical threat was posed, they asserted (Gollan, 2011).

Perhaps more troubling than the sanguine implication that synthetic biology did not represent novel threats was the accompanying attitude that self-regulation had everything under control. Synthetic biologists diligently police themselves, Keasling told the *Times* reporter. But do they? Synberc's administrative director, Kevin Costa, related that security proposals were something officials wanted to implement, they just lacked the leadership to get it done. "We were always interested in doing those things, but there wasn't a champion to step up – whether it be Jay Keasling or Paul Rabinow – and lead those projects," he declared. "I think the bioscientists were looking to Paul to lead the way, and he was waiting for the bioscientists to show more enthusiasm" (Gollan, 2011). Synberc scientists included those from Harvard, Stanford, the Massachusetts Institute of Technology, Prairie View A&M University in Texas, and the University of California, San Francisco. If leadership would not come from one them, from where would it come?

Proponents were not stopping to find out. With synthetic biology advocates securing the federal government's *imprimatur* to self-regulate, and while the Synberc controversy simmered, bioentrepreneurs moved on ambitious plans to create a massive lab compound in the San Francisco Bay Area's East Bay. The Lawrence Berkeley National Lab (LBNL) is part of the U.S. Department of Energy's laboratory system, although the University of California manages it in Berkeley. In January 2011, LBNL (known also as Berkeley Lab) announced plans to launch a second campus. The new complex would combine three labs then scattered throughout the SF East Bay Area: the Joint BioEnergy Institute, the Life Sciences Division, and the Joint Genome Institute (Brownstein, 2011). The unassumingly termed second campus was actually a plan for a sprawling 2 million square foot complex (Brenneman, 2011). LBNL's news release promised that, "Berkeley Lab will seek to co-locate with synergistic commercial and industrial activity to help promote the development of clean and sustainable energy technologies" (Krotz, 2011).

While the campaign to site the lab was underway, an April 2011 news alert from the American Association for the Advancement of Science announced the formation of the UC Berkeley Synthetic Biology Institute (SBI). "SBI will be an important link in a constellation of research centers focused on synthetic biology at UC Berkeley and Lawrence Berkeley National Laboratory (LBNL), both of which have made the field a research priority," it explained. It went on to describe its unique collaboration with "leading companies, designed to translate leading research on biological systems and organisms efficiently into processes, products, and technologies to meet real-world demands" (AAAA, 2011). The stretch from academia to commercial enterprise was seamless and, notwithstanding insufficiently considered hazards, it appeared poised for a towering publicly funded extension into unsuspecting urban communities.

That summer, LBNL's community-relations manager announced it would hold community meetings in each of the six candidate cities it had short-listed. With SBI and Synberc ensconced at the Berkeley campus, surely synthetic biology experiments would be conducted at the planned complex. Would LBNL inform communities? Watchdog groups grew concerned that the cities competing to host the expansion were being sold a gift-wrapped box promoting jobs, increased business, and the prestige of being home to "green technology" research. The tailored packaging seemed familiar. Just as consumables on market shelves bore no labels revealing them to be products of "extreme genetic engineering," an extensive lab complex in a densely populated urban area would mask the processes and risks concealed at its core. When the *Berkeley Daily Planet* featured an article announcing LBNL's plans, it described the lab's work as, "focusing heavily on creating genetically modified organisms, with the labs to be brought onto the new campus focusing in three related areas ... Genomics, Life Sciences, and Physical Biosciences." Citizens and civic groups exhibited no inkling that the modestly dubbed second campus would constitute a global hub for a kind of extreme genetic engineering experimentation that bore inadequately considered hazards. Important questions weren't being asked.

What was the *specific* nature of research to be undertaken at the lab? A grand claim to be working on alternative energy projects could not begin to unpack for local neighborhoods the black box of synthetic biology. Would there be time and opportunity for communities to consider the social and ethical ramifications, locally and globally, of extreme genetic engineering research and development? What were the health and safety risks? Especially in light of the Synberc controversy, it would be important to know: what public safety infrastructure would ensure accountability and transparency?

The question of lab safety alone presented serious questions for local citizens to consider. The issue earned national attention in 2010 when whistleblower Becky McClain won $1.37 million in damages against

Pfizer, Inc. McClain fell seriously ill from what she reported was a genetically engineered virus. When she raised safety concerns with the company, she was fired. The pharmaceutical giant barred her from learning the genetic content of the virus in question because, it said, trade secrecy laws protected it. *New York Times* coverage reported that as some worker advocates viewed it, the case demonstrated "the risks workers in biological labs encounter and the lack of rules to protect them." Consumer advocate Ralph Nader characterized the problem succinctly: "It's a field that has been trade-secreted out of the sunlight" (Pollack and Wilson, 2010).

Laboratories receive biosafety level (BL) rankings according to the risk of harm they represent. They are ranked 1 to 4, the higher the level of danger, the higher the number. Usually, BL1 labs conduct research on non-human infectious agents. Labs with biological agents that might cause "moderate harm" to humans receive a BL2 ranking. Labs working with biological agents that could kill people but for which there is an antidote (e.g., anthrax) are BL3 labs. When there is no known antidote for a biological agent that could kill people, the lab is ranked B4. Would neighborhoods in these densely peopled cities know whether and what pathogens are being used? What did "moderate harm" mean? Moreover, the interconnections between academic and corporate parties presented complex regulatory considerations. How would the regulatory patchwork of different agencies apply when academic and private interests can operate under varying restrictions in terms of safety, liability, and oversight? In the event of lab worker accidents, public safety hazard, or environmental disaster, what remedies would apply? (Gruber et al., 2011).

Activists began organizing to raise public awareness. Institutional organizers or endorsers included Alliance for Humane Biotechnology (with which this book's authors are associated), California BioSafety Alliance, California Coalition for Workers Memorial Day, Center for Environmental Health, Center for Genetics and Society, Center for Food Safety, Council for Responsible Genetics, the ETC Group, Friends of the Earth, Global Justice Ecology Project, Injured Workers National Network, International Center for Technology Assessment, Movement Generation Justice and Ecology Project, and West County Toxics Coalition. Together they formed the ad hoc coalition eventually named SynBioWatch (Jones, 2012). With a grant from the CS Fund/Warsh Mott Legacy, plans to present an informational public forum in the spring of 2012 got underway.

Previously, a dozen civil society groups had sent a letter to the mayors or city councils of each of the seven cities targeted as cites for the new lab (Appendix H). It explained the nature of synthetic biology and recommended that, "before any decisions are made on a specific site for this new lab . . . a comprehensive, independent and transparent safety and risk analysis capable of assessing these [explained] threats must be completed."

It's not clear whether the letter had an impact. When this book's co-author, Tina Stevens, attended a Richmond city community meeting, none there knew what synthetic biology was or whether the new lab would be engaging in such research. Representatives of the City of Albany evidenced some awareness and concern. A November 20, 2011 report from the Albany Waterfront Task Force related that it had been informed that the second campus was "not scheduled to house a UC Berkeley [sic] Synthetic Biology Institute (SBI) in phase 1." But, it noted, LBNL had not announced projects planned for what was to come after. "What types of research does LBNL plan on conducting at the second campus beyond phase one? What are the risks to public health and safety, as well as mitigations and monitoring plans?" it queried. "Information about LBNL's safety and environmental track record is difficult to obtain," it concluded (Albany Report, 2011, no. 6, p. 5). Its December report indicated continuing nondisclosure from the lab. "To date," it related, "no info has been presented on any actual or proposed alternative energy projects for this LBNL campus." It also stated that, "Information about environmental issues related to Lab work [was] unknown" (Albany 2011, no. 7, p. 15).

Activists presented "Unmasking the Bay Area Biolab and Synthetic Biology: Health, Justice, and Communities at Risk" at Berkeley's David Brower Center, on March 29, 2012 (Beitiks, 2012; Rogers, 2012). Local, national, and international presenters offered the standing room only audience a full menu: an explanation of synthetic biology, its unique risks and hazards, what communities needed to know about lab safety, as well as how plans to shift to a biomass economy would adversely affect communities and environments in the global south (Unmasking the Bay Area Biolab, 2012). The city chosen for the second campus had been announced three months earlier: Richmond. Already home to LBNL, Berkeley would have been a convenient choice for the expanded lab complex. But a vocal group of Berkeley citizens, long unsettled by lab expansion into the city, would have made siting the lab in Berkeley a hard sell even were hazards associated with synthetic biology never day-lighted. By contrast, Richmond denizens living in the most impoverished urban area of the region responded to the promise of jobs, job training, and the stimulus for local business. Richmond was also the site of shoreline property already owned by UC Berkeley, the Richmond Field Station (RFS).

On January 23, 2012, a press conference held at the RFS, described as "crowded with jubilant City of Richmond officials," announced Richmond as the pick. The doors of the new campus, forecasted the LBNL director, would open in 2016 (Ness, 2012). The very next day, a local CBS news source revealed what the City of Albany seemed not to know as late as the preceding month, "A Lawrence Berkeley Lab scientist," it reported, "claims a new research campus in Richmond will work on reducing U.S.

dependence on foreign oil by turning bio-mass into fuel." It quoted Jay Keasling, who was not identified as Director of Synberc or as co-founder of Amyris Biotechnologies, a bioenergy startup. In 2010, the company's public offering valued his one million shares at approximately $17 million (Abate, 2010). "I represent the more than 800 people in biology and bio-engineering at Lawrence Berkeley National Laboratory," Keasling declared, "who will be the first people to inhabit this site, and do work at this site" (CBS, 2012). The boast would not be borne out. The forecasted groundbreaking never materialized.

Media sources reported in September 2013 that the lab had run into funding problems (Leuty, 2013; Matier and Ross, 2013; Tucker, 2016). Within a year and eight months, it had lost expected funding from the Department of Energy owing, it claimed, to federal sequestration. The following year, UCB announced plans to repurpose the RFS. Its sketch suggested that visions of a lab complex had vanished. Instead, the site would be home to the "Berkeley Global Campus" and UCB would "become the first American university to establish an international campus in the United States." It would offer, "a curriculum centered on global governance, ethics and political economy, cultural and international relations" (*Berkeley News*, 2014). This too, however, would prove a pipe dream (Melles and Shea, 2016). In February 2015, the *LA Times* reported that, "UC officials acknowledge that the proposal faces many hurdles, especially in undetermined funding that could reach into the billions" (Gordon, 2015). By August 2016, local papers reported that UCB was suspending plans altogether for the 34-acre, multi-billion dollar global campus owing to "budgetary challenges" (Taylor, 2016; Rauber, 2016).

The federal government was not out of the business of dispensing largesse to synthetic biologists, however. While it did not fund the LBNL lab park, it offered lavish funding for a more modest project in San Francisco. In October 2016, the *San Francisco Chronicle* reported that a National Science Foundation grant of $25 million would fund a "Center for Biological Construction" (CBC). Although there would be no new laboratories requiring civic approval, the announcement did not downplay the project's focus. The CBC was meant to be a consortium dedicated to synthetic biology research and development. Headed by scientists at the University of California at San Francisco, it would include specialists from Stanford, UC Berkeley, San Francisco State, and the artificial intelligence lab at the IBM-Almaden research center in San José. Although the center's leaders would be located in San Francisco, they would continue to conduct research at their individual labs. And, university researchers, according to UCSF, "will work with their counterparts in commercial companies" (Perlman, 2016). The report, as is typical of journalistic coverage of these fields, left unaddressed whether or how many of the grantees were either founders or on the boards of biotech corporations.

Coda

When activists decided to launch a public forum on synthetic biology and the prospect of a new lab compound for San Francisco's East Bay, they recognized attributes of a juggernaut in action. It was a long list: 1) how unforthcoming the Lawrence Berkeley National Lab (LBNL) and UC Berkeley (UCB) were about the specific nature of the research expected to take place at the proposed lab compound; 2) how powerful the unsupported promises of jobs and "green-washed" (i.e., having unsupported environmental benefits) technology were in beguiling candidate cities; 3) how easy it was for bioentrepreneurs to trade on their credentials as university researchers, escaping public recognition of corporate ties and avoiding discussion of how government largess for science funding can bend toward corporate welfare; 4) what little consideration was given for providing community awareness of and civic guidance over hazards associated with extreme genetic engineering; 5) the lack of attention synthetic biologists gave to the human costs of what they were doing for communities around the globe.

Moreover, underlying the reactions of many critics was the conviction that the synthetic biology methods being devised, once worked out in nonhuman systems, would someday be turned to the modification of our own species. Certainly, this is the stated intention of a number of those spearheading the field. Concerns were borne out when, in May 2016, synthetic biology advocates wanting to synthesize a human genome from scratch met behind closed doors at Harvard University (Endy and Zoloth, 2016). Chapter 6, "The Road to Gattaca," examines the world that bioentrepreneurs and their allies in newly conceived areas of medicine are attempting to forge from these ingredients and techniques.

Notes

1 Portions of Chapter 5 were adapted from Newman, S.A. (2012), "Synthetic Biology: Life as App Store," *Capitalism Nature Socialism* 23, 6–18.
2 The implications of combining these technologies for improving humans to gain commercial and military advantage were explored almost two decades ago in a 2002 joint report by the U.S. National Science Foundation and Department of Commerce ("Converging Technologies for Improving Human Performance: Nanotechnology, Biotechnology, Information Technology and Cognitive Science," www.wtec.org/ConvergingTechnologies/Report/NBIC_report.pdf). The convergence of the technologies was discussed critically in 2015 by the Center for Genetics and Society (Berkeley, CA), and Friends of the Earth (Washington, D.C. and Berkeley) in a joint report, "Extreme Genetic Engineering and the Human Future: Reclaiming Emerging Biotechnologies for the Common Good," https://1bps6437gg8c169i0y1drtgz-wpengine.netdna-ssl.com/wp-content/uploads/wpallimport/files/archive/FOE_ExtremeGenEngineering_10.pdf

Sources Consulted for Chapter 5

AAAA, "UC Berkeley Launches Synthetic Biology Institute to Advance Research in Biological Engineering," *EurekAlert!*, April 21, 2011: www.eurekalert.org/pub_releases/2011-04/pmac-ubl041911.php

Abate, Tom, "Jay Keasling Hits Jackpot with Biofuel Startup," *SF Gate*, November 21, 2010: www.sfgate.com/news/article/Jay-Keasling-hits-jackpot-with-biofuel-startup-3245230.php

Albany Waterfront Task Force, Meeting No. 6, November 20, 2011: http://static1.1.sqspcdn.com/static/f/960884/15190968/1321644857097/Nov-20-Final-TaskForce.pdf

Albany Waterfront Task Force, Meeting No. 7, December 4, 2011: www.albanyca.org/home/showdocument?id=17706

Auslander, S., M. Wieland, and M. Fussenegger, "Smart Medication through Combination of Synthetic Biology and Cell Microencapsulation," *Metab Eng* 14 (2011): 252–260.

Ball, Philip, "Synthetic Biology: Starting from Scratch," *Nature* 431 (October 7, 2004): 624–626: www.nature.com/articles/431624a

Baumgaertner, Emily, "As D.I.Y. Gene Editing Gains Popularity, 'Someone Is Going to Get Hurt'," *New York Times*, May 14, 2018: www.nytimes.com/2018/05/14/science/biohackers-gene-editing-virus.html

Beitiks, Emily Smith, "Bay Area May Be at Risk from Synthetic Biology Research Labs," *The Mercury News*, March 27, 2012: www.mercurynews.com/2012/03/27/emily-smith-beitiks-bay-area-may-be-at-risk-from-synthetic-biology-research-labs/

Berkeley News, "Berkeley Global Campus: A New Bolder Vision for Richmond Bay," October 30, 2014: http://news.berkeley.edu/2014/10/30/berkeley-global-campus/

Biofuelwatch, "New Report Argues Algae Biofuels Are Overhyped and Unacceptably Risky," *Biotechnology for Biofuels*, September 26, 2017: www.biofuelwatch.org.uk/2017/report-microalgae-biofuels-overhyped-pose-risks-to-ecosystems-and-public-health/

Bohannon, J., "The Life Hacker," *Science* 333 (2011): 1236–1237.

Brenneman, Richard, "Lawrence Berkeley Lab's RFQ Points toward Richmond Site Choice," *Berkeley Daily Planet*, January 4, 2011: www.berkeleydailyplanet.com/issue/2011-01-05/article/37066

Brownstein, Zelda, "LBNL Announces Community Meetings on Second Campus at Berkeley Chamber Forum," *Berkeley Daily Planet*, June 8, 2011: www.berkeleydailyplanet.com/issue/2011-06-08/article/37958?headline=LBNL-Announces-Community-Meetings-on-Second-Campus-at-Berkeley-Chamber-Forum-News-Analysis-

CBS SF Bay Area, News Item: "New Richmond Lab Expected to Create Innovative Bio-Fuel," January 24, 2012: http://sanfrancisco.cbslocal.com/2012/01/24/new-richmond-lab-expected-to-create-innovative-bio-fuel/

Center for Genetics and Society, Friends of the Earth, "Extreme Genetic Engineering and the Human Future: Reclaiming Emerging Biotechnologies for the Common Good," 2015: www.geneticsandsociety.org/sites/default/files/human_future.pdf

Civil Society Letter to President's Commission on Synthetic Biology: www.etcgroup.org/sites/www.etcgroup.org/files/publication/pdf_file/Civil%20Society%20Letter%20to%20Presidents%20Commission%20on%20Synthetic%20Biology_0.pdf

Constante, M., R. Grunberg, and M. Isalan, "A Biobrick Library for Cloning Custom Eukaryotic Plasmids," *PLoS One* 6 (2011): e23685.

Council for Responsible Genetics, "Biowarfare and BioLab Safety," n.d.: www.councilforresponsiblegenetics.org/Projects/PastProject.aspx?projectId=4

Daily Californian, Editorial, "A Fond Farewell to the Berkeley Global Campus," August 30, 2016: www.dailycal.org/2016/08/30/fond-farewell-berkeley-global-campus/

Endy, Drew, and Laurie Zoloth, "Should We Synthesize a Human Genome," DSpace@MIT, May 10, 2016: https://dspace.mit.edu/handle/1721.1/102449

ETC Group, "Extreme Genetic Engineering: An Introduction to Synthetic Biology," January 16, 2007: www.etcgroup.org/content/extreme-genetic-engineering-introduction-synthetic-biology

ETC Group, "Groups Criticize Presidential Commission's Recommendations on Synthetic Biology," December 16, 2010: www.etcgroup.org/content/groups-criticize-presidential-commission's-recommendations-synthetic-biology

ETC Group, "What Is Synthetic Biology: Engineering Life and Livelihoods," October 29, 2014: www.etcgroup.org/sites/www.etcgroup.org/files/files/synbio_comics-complete_letter_size_rev.pdf; video at: www.synbiowatch.org/2014/10/synthetic_biology_explained/?lores

Friends of the Earth, "Synthetic Biology 101," n.d.: https://foe.org/resources/synthetic-biology-101/

Friends of the Earth, "Principles for the Oversight of Synthetic Biology," March 13, 2012: https://foe.org/2012-03-global-coalition-calls-oversight-synthetic-biology/

Friends of the Earth, "Biotech Industry Cooks Up PR Plans to Get Us to Swallow Synthetic Biology Food," May 22, 2014: https://foe.org/2014-05-the-synthetic-biology-industrys-pr-scheme/

Gollan, Jennifer, "Lab Fight Raises U.S. Security Issues," New York Times, October 22, 2011: www.nytimes.com/2011/10/23/us/synberc-fight-raises-national-security-issues.html

Gordon, Larry, "UC Berkeley Studies International Education Campus in Richmond," LA Times, February 24, 2015: www.latimes.com/local/education/la-me-uc-richmond-20150224-story.html

Gruber, Jeremy, Tina Stevens, and Becky McClain, "A Synthetic Biology Lab in Berkeley," GeneWatch 24(3) (June–July 2011): www.synbiowatch.org/wp-content/uploads/2012/01/GeneWatch-Biolab-correct.pdf

Hayden, E.C., "Bioengineers Debate Use of Military Money," Nature 479 (2011): 458.

Haynes, K.A., and P.A. Silver, "Synthetic Reversal of Epigenetic Silencing," J Biol Chem 286 (2011): 27176–27182.

Heuer, H., and K. Smalla, "Horizontal Gene Transfer Between Bacteria," Environmental Biosafety Research 6 (2007): 3–13.

Hoffman, Eric, and Stuart A. Newman, "Big Promises Backed by Bad Theory," GEN: Genetic Engineering and Biotechnology News," May 15, 2012: www.genengnews.com/gen-articles/big-promises-backed-by-bad-theory/4104/

Huang, Y.C., B. Ge, D. Sen, and H.Z. Yu, "Immobilized DNA Switches as Electronic Sensors for Picomolar Detection of Plasma Proteins," Journal of the American Chemical Society 130(25) (2008): 8023–8029.

Isaacson, Walter, Steve Jobs, New York: Simon & Schuster, 2011.

Jones, Stephen T., "Playing God?," San Francisco Bay Guardian, in Guardian Archive 46(27) (2011–2012, April 3, 2012): http://sfbgarchive.48hills.org/sfbgarchive/2012/04/03/playing-god/

Kean, S., "A Lab of Their Own," Science 333 (2011): 1240–1241.

Krotz, Dan, "Lawrence Berkeley National Laboratory Launches Information Gathering Process for a Possible Second Campus," News Release, Berkeley Lab: http://newscenter.lbl.gov/2011/01/03/second-campus/

Kunjapur, A., "An Introduction to Start-Ups in Synthetic Biology," Phys Org (2015): https://phys.org/news/2015-09-introduction-start-ups-synthetic-biology.html

Leuty, Ron, "Future of Berkeley Lab's Massive Second Campus 'Uncertain'," San Francisco Business Times, September 16, 2013: www.bizjournals.com/sanfrancisco/blog/biotech/2013/09/uc-berkeley-lawrence-lab-richmond.html

Lloyd's, "Synthetic Biology: Influencing Development," Lloyd's Emerging Risks Team Report, July 2009, Version 1: www.lloyds.com/news-and-risk-insight/risk-reports/library/technology/synthetic-biology

McFadden, J., "It's Time to Play God," The Guardian, August 23, 2009: www.theguardian.com/commentisfree/2009/aug/23/venter-artificial-life-genetics

MarketWatch, "Synthetic Biology Market Is Forecast to Cross US$ 38 Billion By 2020," April 17, 2018: www.marketwatch.com/story/synthetic-biology-market-is-forecast-to-cross-us-38-billion-by-2020-2018-04-17

Matier, P., and A. Ross, "Berkeley Lab's Contract Loss Threatens Richmond Expansion," San Francisco Gate, September 16, 2013: www.sfgate.com/bayarea/matier-ross/article/Berkeley-lab-s-contract-loss-threatens-Richmond-4816913.php

Melles, Nebiat Assefa, and Jacob Shea, "Richmond Reacts to Berkeley Global Campus Suspension," September 2, 2016: http://richmondconfidential.org/2016/09/02/richmond-reacts-to-berkeley-global-campus-suspension/

Nandagopal, N., and M.B. Elowitz, "Synthetic Biology: Integrated Gene Circuits," Science 333 (2011): 1244–1248.

Nature, editorial, "Futures of Artificial Life," 431, 613 (October 7, 2004): www.nature.com/articles/431613b

Ness, Carol, "Lab Picks Richmond Field Station for a Second Campus," UC Berkeley Media Release, January 23, 2012: https://vcresearch.berkeley.edu/news/lab-picks-richmond-field-station-second-campus

New Directions: The Ethics of Synthetic Biology and Emerging Technologies. 2010. Report of the Presidential Commission for the Study of Bioethical Issues. December. https://bioethicsarchive.georgetown.edu/pcsbi/sites/default/files/PCSBI-Synthetic-Biology-Report-12.16.10_0.pdf

Newman, Stuart A., "Meiogenics: Synthetic Biology Meets Transhumanism," *GeneWatch* 25(1–2) (2006): www.synbiowatch.org/wp-content/uploads/2013/05/Genewatch_Meiogenics.pdf

Newman, Stuart A., "Synthetic Biology: Life as App Store," *Capitalism Nature Socialism* 23(1) (March 2012): 6–18.

Newman, Stuart A., "The Demise of the Gene," *Capitalism Nature Socialism* 24(1) (2013): 62–72.

Open Letter to President's Bioethics Commission from Fifty-Eight Civil Society Groups, December 16, 2010: www.etcgroup.org/sites/www.etcgroup.org/files/publication/pdf_file/Civil%20Society%20Letter%20to%,20Presidents%20Commission%20on%20Synthetic%20Biology_0.pdf

Perlman, David, "Bay Area Cellular Machine Shop to Be Created with Federal Grant," *San Francisco Chronicle*, October 2, 2016: www.sfchronicle.com/science/article/Bay-Area-Cellular-Machine-Shop-to-be-created-with-9537679.php?ipid=gsa-sfgate-result%3fcmpid=email-premium

Pollack, Andrew, "U.S. Bioethics Commission Gives Green Light to Synthetic Biology," *New York Times*, December 16, 2010: www.nytimes.com/2010/12/16/science/16synthetic.html?_r=1

Pollack, Andrew, and Duff Wilson, "A Pfizer Whistle-Blower Is Awarded $1.4 Million," *New York Times*, April 2, 2010: www.nytimes.com/2010/04/03/business/03pfizer.html

Rabinow, Paul, and Gaymon Benett, *Designing Human Practices: An Experiment with Synthetic Biology*, Chicago, IL: University of Chicago Press, 2012.

Rauber, Chris, "UC Berkeley Backs Off on Plans for Big New Campus in Richmond," *San Francisco Business Times*, August 26, 2016: www.bizjournals.com/sanfrancisco/news/2016/08/26/uc-berkeley-richmond.html

Rogers, Robert, "Critics Raise Safety Concerns with Biotech Labs at Berkeley Forum," *East Bay Times*, March 28, 2012: www.eastbaytimes.com/2012/03/28/critics-raise-safety-concerns-with-biotech-labs-at-berkeley-forum/

Sharp, Daniel, "Craig Venter in Space: Houston We Have a Problem," *Biopolitical Times*, April 5, 2012: www.geneticsandsociety.org/biopolitical-times/craig-venter-space-houston-we-have-problem?id=6151

Skerker, J.M., J.B. Lucks, and A.P. Arkin, "Evolution, Ecology and the Engineered Organism: Lessons for Synthetic Biology," *Genome Biol* 10(11) (2009): 114.

Specter, M., "A Life of Its Own: Where Will Synthetic Biology Lead Us?," *The New Yorker*, September 28, 2009, pp. 56–65.

Synbiobeta, "Synthetic Biology Companies Raised Over $650 Million in Q1, Setting the Pace for Another Record-Breaking Year," March 31, 2018: https://synbiobeta.com/synthetic-biology-companies-raised-over-650-million-in-q1/

Taylor, Tracy, "UC Berkeley Suspends Plans for Richmond 'Global Campus'," *Berkeleyside*, August 26, 2016: www.berkeleyside.com/2016/08/26/uc-berkeley-suspends-plans-for-richmond-global-campus

Thomas, Jim, "Long Now Synthetic Biology Debate," *Wired Science*, 2008, video at: www.youtube.com/watch?time_continue=10&v=gDQXUR0Pb8c

Tucker, Jill, "UC Berkeley Suspends Plans to Build Global Campus in Richmond," *San Francisco Gate*, August 26, 2016: www.sfgate.com/bayarea/article/UC-Berkeley-suspends-plans-to-build-Global-Campus-9187281.php

Unmasking the Bay Area BioLab and Synthetic Biology, March 29, 2012: http://sanfrancisco.eventful.com/events/berkeley-conference-unmasking-bay-area-biolab-and-/E0-001-046869647-9

Woodrow Wilson Center, Synthetic Biology Project: www.synbioproject.org/cpi/

6
The Road to Gattaca[1]

Many people love their retrievers and their sunny dispositions around children and adults. Could people be chosen in the same way? Would it be so terrible to allow parents to at least aim for a certain type, in the same way that great breeders . . . try to match a breed of dog to the needs of a family?

(Gregory Pence, Philosophy Professor,
Alabama, 1998, p. 168)

The GenRich . . . carry synthetic genes. All aspects of the economy, the media, the entertainment industry, and the knowledge industry are controlled by . . . the GenRich class . . .. Naturals work as low-paid service providers or as laborers. [Eventually] the GenRich class and the Natural class will become entirely separate species . . . with as much romantic interest in each other as a current human would have for a chimpanzee.

(Lee Silver, bioentrepreneur, Professor,
Molecular Biology, Princeton University, 1997, p. 6)

The mingling of individual human chromosomes with those of other mammals assures a gradualistic enlargement of the field . . .. the issue of "subhuman" hybrids may arise first, just because of the touchiness of experimentation of obviously human material.

(Joshua Lederberg Nobel Laureate, Stanford
molecular biologist from, "Experimental
Genetics and Human Evolution," 1966)

For as long as he could remember, Vincent Freeman dreamed of becoming an astronaut. He wasn't sure why. It could have been his love of distant planets, he once mused. More sharply, he suggested that maybe it was his growing dislike of life on Earth. But one thing always was clear to him: space travel was not a passing passion, it was an undying desire. It was also an ambition that his parents, loving him, persistently discouraged.

"The only way you'll see the inside of a spaceship is if you're cleaning it," his Dad warned brutally. The caution struck like a stinging slap. On some level, though, Vincent understood that his Dad was not wrong. It was no secret that although conceived in love, his parents had conceived Vincent naturally – without genetic interventions. So, the nature of his conception left him vulnerable to social condemnation for possessing even the *possibility* of developing conditions such as premature baldness or obesity. Certainly, he possessed none of the genetic enhancements that reproductive bioscience made available to prospective parents. And in fact, after he was born, a genetic test revealed a real chance (though not an inevitability) of his developing a potentially fatal heart condition. Vincent was a "de-gene-erate," an "in-valid." He was part of a new underclass, one not based on social class or skin color. As he understood it, his unwilling membership in this underclass was a twisted function of society getting discrimination "down to a science."

Vincent Freeman is the protagonist of the 1997 dystopian film, *Gattaca*. For over two decades, commentators have valued the film as an apt thought-exercise offering deep understanding of what a society in the full embrace of techno-eugenics looks like. Center for Genetics and Society co-founder Richard Hayes coined the term "techno-eugenics" to short-hand the social effects of the merging of genetic and reproductive technologies within the context of a consumer driven market (Hayes, 2000). This post-Second World War confluence reinvigorated discredited nineteenth and early twentieth century impulses to breed "better" humans and eliminate "inferior" ones. The intimate portrayal of Vincent's world is a social portrait that ably teases out only dimly recognized implications of these historic forces. It is a depiction finding intensifying relevance as the clutch of techno-eugenics tightens.

In the film discrimination against in-valids, "genoism," is illegal. But Vincent knows first-hand that no one takes that law seriously. How else, he wryly wonders, did it come to pass that as an adult, he had cleaned half the toilets in the state? "No matter how much I trained or how much I studied," Vincent explains, "the best test score in the world wouldn't matter unless I had the blood test to go with it." His resume, he scoffs, "was in his cells." The apparent self-mockery only thinly masks a snarling social critique: the market in human genes had encoded prejudice and injustice into the human genome (*Gattaca*, 1997).

The film's title references the name of the fictive company that launches missiles into space. The four letters forming the company's name, G A T C, represent the four nucleotides, Guanine, Adenine, Thymine, and Cytosine that, in varying combinations, form the DNA molecule. In a techno-eugenic society, DNA functions as gatekeeper. Vincent must secure employment at Gattaca if he is ever to realize his dream. Without the right genes, Vincent does not qualify for that employment. Janitor? Detective? Astronaut? Genetic prejudices at Gattaca set qualifications and dictate careers. DNA is a gatekeeper for even more, of course. Romance, for example. A strand of hair, snatched surreptitiously or voluntarily offered-up for genetic sequencing, can tell a love interest whether your genes are worth dating. Had your parents chosen genetic traits wisely? Had they been constrained to choose affordably? What genetic choices were available the year of your creation? A suitor might suppress early amorous impulses and pass over a love interest. Deny kismet. Hold out for someone spawned from an upgraded enhancement menu. Hold out for Humans 3.0, if you can.

The film's conclusion, which sees Vincent boarding for space flight, serves up conflicting moral reflections. The ways in which the film's genetically enhanced humans suffer or fail or function as scofflaws to assist Vincent in gaming Gattaca's system demonstrates the burdens and limitations of a society choked by stultifying standards of flawlessness. And when Vincent ultimately succeeds it is understood that his journey along the long arc of his rocket's ascent had been mapped by the creative vitality of unmodified human spirit. But the subtext is grim. The world from which Vincent escapes remains marred. There, the mere *perception* that molecular enhancements can improve people, whether or not "improvement" ever actually results, leaves society debased, scarred irredeemably by genetic castes (Darnovsky, 2016). Every caste level – the genetic have-nots, the genetic-haves, the genetic haves of next-generation enhancements – reeks of contempt and self-loathing. As one commentator concludes,

> the film reveals how individual choices made in pursuit of human perfection can lead to a different type of coercion that exceeds what any law can mandate. Parents can come to feel that they don't really have a choice if they want their children to be successful, leading those who are screened for greatness to ultimately become imprisoned by others' expectations.
>
> (Obasogie, 2017)

Vincent Freeman, despite his caste, is the film's only free man.

The film is one that creators of genetic technologies seem not to want to talk about. A 2015 *Nature* editorial specifically advised its professional readership not to engage discussion of the film. "When controversy comes calling," it counseled, "scientists should reach outwards. Discussions should ... include ... the broader public. ... [and] should avoid unhelpful references to the genetically modified humans in the 1997 film *Gattaca*." (The *New York Times* science writer Carl Zimmer went so far as to call on his personal blog site for "an international ban on invoking *GATTACA* in these discussions" (cited in Darnovsky, 2015).)

Sometimes, reportage reassures that science is not yet capable of creating "designer babies." The online journal, *STAT*, for example, headlined in the summer of 2017 that, "U.S. scientists edit genome of human embryo, but cast doubt on possibility of 'designer babies'." It quotes a researcher declaring that, "more research would be needed to really know how efficiently a new gene version can be introduced" (Begley, 2017a). Such qualms seem genuine. But when the dam burst in late 2018 with news of the purported gene engineering of twin girls in China, some key U.S. players saw an opportunity to speed things up (Cohen, 2018; Kelly, 2018). As described below, this readiness to leap was in the cards for more than two decades. Journalists and bioethicists, unable or unwilling to secure access to better information had been conducting, perhaps unwittingly, exercises in public relations.

Paving the Road to Gattaca

In 2015, *Science Magazine* hailed CRISPR as a breakthrough technology. The catchy acronym stands for a more cumbersome descriptor. Microbiologists in 2001, recognizing the significance of a collection of repetitive DNA sequences, played with securing for it a "snappy" sounding neologism. They rejected the descriptor, "spacers interspersed direct repeats" which would have yielded the dismaying sounding acronym, "Spidr." Instead, they chose, "clustered regularly interspaced short palindromic repeats," resulting in the more congenial "CRISPR." Its sometimes employed suffix, "Cas" with an accompanying number, in this case the number 9, stands for the enzymes that act on the CRISPR DNA sequences to do the desired cutting and pasting. When, 14 years later, the acronym CRISPR-Cas9 leapt out of the lab, the abbreviated "CRISPR" proved an effective mnemonic device for popularizing the technology (Zimmer, 2016). It is also proving effective in normalizing it. By July 2017, when a *Washington Post* article queried, "If you could 'design' your own child, would you?" it proclaimed the "era of human gene editing has begun," and concluded that, "CRISPR's seductiveness is beginning to overtake the calls for caution" (Wadhwa, 2017).

CRISPR is, indeed, a technological game-changer. Recombinant DNA gene splicing techniques of the previous three decades offered only hit-or-miss gene insertions and replacements. Modification of one or a few base pairs was, essentially, impossible. CRISPR made these more precise "edits" feasible, if not foolproof. It was also simple to perform, and cheap. But the technology, described often as "revolutionary," was proving unsettling even for some of its creators, despite the whimsical origins of its name or how swiftly it settled into the public lexicon.

A few years after she and her colleagues published a paper in 2012 describing the gene editing uses of CRISPR, UC Berkeley's Jennifer Doudna, began having trouble sleeping. "By the spring of 2014," she later recounted,

"I was regularly lying awake at night wondering whether I could justifiably stay out of an ethical storm that was brewing around a technology I had helped to create." With astonishing speed, CRISPR had become a widely employed lab tool used to alter the genomes of a wide variety of plants and animals easily and with unprecedented, if not perfect, precision. How long would it be, she wondered, before someone tried it in humans? She saw it as inevitable that, "researchers somewhere would test the technique in human eggs, sperm or embryos, with a view to creating heritable alterations in people." Doudna's fitful nights led her to help convene a one-day conference in Napa, California in January 2015. Its 18 attendees, mostly scientists and bioethicists, published a statement in April of that year urging the "scientific community" to refrain – at this time – from using gene editing tools to modify human embryos for clinical (i.e., intended for full-term birth) applications. They also called for further public meetings to discuss responsible pursuit of genome engineering research and application.

One of her motivations in calling for the 2015 Napa conference, Doudna relates, was to address her concern that the technology "would be used in a way that was either dangerous, or perceived to be dangerous, before scientists had communicated enough about it to the wider world. " Biology students should be trained, she urged, to speak to non-scientists. Future researchers should learn how to create an effective "elevator pitch" to describe their work (Doudna, 2015). But by 2016 she told a reporter for the *Wall Street Journal* that she anticipates learning about the first baby whose genes have been altered in a lab within the next ten years, or even sooner (Wolfe, 2016). By January 2018, she explained that her views had "evolved." Now, she declared, "it may be worth doing" (UC Berkeley Events, 2018).

Since the Napa meeting, there have been many hearings and global "summits," some public, some behind closed doors. Have they functioned to broaden public understanding or, for that matter, input?

The most far-reaching of the public convenings took place in December 2015. Washington D.C. was the site of the three-day, International Summit on Human Gene Editing. The U.S. National Academy of Sciences, the U.S. National Academy of Medicine, the Chinese Academy of Sciences, and the United Kingdom's Royal Society co-sponsored the event. At its conclusion, the sponsors' statement declared that it would be irresponsible to proceed with clinical use of germline editing without "broad societal consensus" (NAS, 2015). But a scant 15 months later saw this precautionary recommendation upended. In February 2017, a committee of the U.S. National Academy of Sciences and U.S. National Academy of Medicine declared that, "clinical trials using heritable germline genome editing should be permitted" (NAS, 2017). Broad social consensus, as it turns out, would no longer be required.

In July 2018, a British organization, with far-reaching influence in the US, pushed aside unresolved ethical considerations to authorize moving closer to human genetic engineering. The Nuffield Council is a semi-official UK bioethics agency. It endorsed stepping up to germline modification declaring, with approval, that "[t]here is potential for heritable genome editing interventions to be used at some point in the future in assisted human reproduction, as a means for people to secure certain characteristics in their children" (Sample, 2018). It made this decision with no new evidence that the hazards of tinkering with genes at early stages of development were anywhere near solved.

The NAS and Nuffield positions fly in the face of what arguably *does* constitute a social consensus – a global consensus against human germline modification. The Council of Europe's 1997 Oviedo Convention, UNESCO's 2015 updated Report on the Human Genome and Human Rights, and over 40 countries have taken positions banning the genetic alteration of human gametes or embryos for reproduction (Lowthorp, 2017). In spite of such resistance, bioentrepreneurs continue to push boundaries. Projects involving genetic manipulation of human embryos employing CRISPR and other methods are well underway. The long-term ramifications of these projects for human safety, social justice, and the evolution of the species are profound and irreversible. Yet, despite their deep biological and social consequences, projects advance without broad public awareness, understanding, or significant democratic input.

A 2015 *Nature* multi-authored editorial bore the title, "Don't Edit the Human Germ Line" (Lanphier et al., 2015). The authors were not themselves engaged in germline editing but in commercially developing therapeutics that could be derived from genome editing of somatic (non-reproductive) cells. "We are concerned," they asserted, "that a public outcry about such an ethical breach [i.e., human germline editing] could hinder a promising area of therapeutic development, namely making genetic changes that cannot be inherited." In the current U.S. political climate, perhaps the only pushback against human germline modification with any potential of engaging the regulatory process may be from commercial interests focused on postnatal stem cell-based treatments understandably skittish about radical procedures for which highly publicized failures could bring down the entire industry.

As pitfalls on the road to Gattaca come into focus, hazards unrecognized or only tangentially explored by the film come into bold relief. They yield unsettling questions consistently minimized or completely ignored by industry-led discussion: How does society regard the producers of the raw materials required by these technologies, namely young egg providers? Do egg providers always understand how their eggs are being used – to assist in reproduction?, for research?, what kind of research? Are the long-term health effects of egg harvesting adequately understood? How do the eugenic impulses driving the march to Gattaca value or disregard the civil rights and social critique of people with disabilities? (Since the 1970s, disability-justice activists succeeded in redefining what it means to be dis-abled. Previously, viewed solely as a pathology located within people considered disabled, the disability rights movement shifted attention to the ways in which society dis-ables individuals. Solutions should not be about "fixing" individuals, but about repairing society (cf. Longmore Institute video).) What are the consequences of ignoring mounting evidence challenging the genetic determinist assumptions underlying embryo modifying techniques? How will society regard those laboratory-created individuals who exhibit "errors" months or years after birth? Will we recognize the source or cause of developmental problems or track individuals to detect them? If current behavior regarding present-day egg providers is any indicator, the future on this score is bleak. Over the many decades that human eggs have been in demand for laboratory embryo creation, the reproductive and research industries have ignored calls for robust evaluation of the long-term health of the young women who provide eggs (see Chapters 3 and 4).

Questions such as these are either wholly ignored or skillfully side-stepped by organizers of meetings convened to consider the social ramifications of genetically manipulating human embryos. And yet, answers to these questions are the dots that need connecting if Gattaca is to be avoided. Why do bioscientists, bioentrepreneurs, and mainstream bioethicists turn away from their examination? How, instead, do they describe the reasons for engaging in activities that virtually ensure arrival at Gattaca? The road to that destination is demarcated by two tracks of bioentrepreneurial behavior. The first is the promotion of a search for cures as the driver of biotech research and development, thereby masking other forces such as commercial incentives and professional advancement. The second is failing either to recognize or accept the fallacy of genetic reductionism.

Bioentrepreneurial conflicts of interest make it difficult for neutral parties to assess whether or not a particular technology is the best way to secure a particular cure. The first sources of whatever the general public can know or understand about a genetic technology are those who are developing it. Often, those standing to profit from use of a genetic engineering technique are those pushing it forward. Hotly contested patent disputes concerning who owns the CRISPR technique and the rights to license it for treatments and drug design make evident the influence of commercial opportunities (Ledford, 2017). Invariably, however, the justification offered by boosters is that genetically modifying embryos is the best way to eliminate inherited disabling conditions.

Some of these conditions, such as Tay-Sachs disease and Type A Niemann-Pick disease, are invariably fatal. But others are variable in their expression and increasingly treatable, such as sickle cell disease and cystic fibrosis. Still others, like congenital deafness or Down syndrome, are not even universally perceived as undesirable. Despite the difference among diseases, however, the goal of preventing "defective" people is generally unquestioned by biotechnology promoters. Moreover, only reluctantly do backers acknowledge the existence of other methods that make it possible to have children free of certain gene-related conditions. Pre-implantation genetic diagnosis (PGD) allows for the implantation of embryos free of the genetic markers of diseases for which they are tested. While not without its own ethical difficulties, it is far less risky compared with germline or other embryo-stage gene modifications to the health of future children, and to the future landscape of social equality and fairness. But a media analysis by the Center for Genetics and Society found that during an 18 month period between 2015 and 2016, only 6 out of 40 articles or commentaries even mentioned the existence of PGD. Media coverage and commentary often give the erroneous impression that imminent cures will be found for millions (CGS, September 2017) – even though engineering people who don't yet exist is definitely not what medicine has traditionally characterized as "cures."

Bioentrepreneur Shoukrat Mitalipov is President of Mitogenome Therapeutics, Inc., and a researcher at Oregon Health and Research University. He appears to have little problem with creating full-term genetically engineered humans. A developer of the controversial "three-parent embryo" genetic engineering technology (see pp. 130ff.), Mitalipov raised eyebrows in 2015 when he applied to expand clinical use of that technology as a treatment for infertility (Connor, 2015). By mid-2018 a commercial venture, Darwin Life-Nadiya, with just this business plan was in place in Ukraine under the direction of U.S.-based Dr. John Zhang, and several babies had already been produced using pronuclear transfer, a three-person technique designed to correct mitochondrial defects (see p. 133). The infertility of the prospective parents who were clients of this clinic, however, had no demonstrated connection to mitochondrial deficiency (Stein, 2018).

In 2017, Mitalipov bounded over yet another previously uncrossed red line, fending off anxieties over having done so with the boast of moving toward a cure. That summer, he announced that his lab had successfully used CRISPR technology to "edit" the DNA of human embryos. The federal government may choose not to fund human embryo gene editing research. But private funders get and give a green light. Unconstrained by a requirement of prior public awareness or approval, he had breached the boundary easily. When he did so, he took all the world, including those unaware and unconsenting, further down the road to Gattaca.

In the wake of the announcement, volleys between promoters and critics formed familiar narrative arcs. Enthusiasts hailed the move as a way to correct a potentially fatal heart condition. Critics noted that methods for eliminating the disease from the gene pool already existed – methods that don't widen the door to genetically modifying humans. General reportage tended to ignore controversies associated with Mitalipov's research and development: his work cloning human embryos, his engineering of human embryos with DNA from three people, his collaboration with the infamous discredited cloning and stem cell fraudster Hwang Woo-suk, or his interest in commercializing techniques he developed in fertility clinics (CGS, July 2017). And only in passing fashion did a reporter acknowledge that, "Human eggs are the key starting point for the groundbreaking experiments." Indeed, Mitalipov has access to eggs from a fertility clinic in the same building that houses his lab (Stein, August 2017).

The powerful commercial incentives to advance genetic technologies stretching along one side of the road to Gattaca are paralleled on the other by a second feature typical of bioentrepreneurial behavior: failure either to recognize or accept the demise of genetic determinism. Genetic determinism is a theoretical paradigm purporting that the role of genes in determining traits is straightforward and predictable. It constitutes the teetering undergirding of those projects seeking clinical use of genetically engineered human embryos. In terms of the long-term health of the full-term humans they result in, how successful genetically engineered embryos end up being depends profoundly on our understanding of gene action, which is increasingly unstable and fraught with exceptions.

If genes cannot be deployed or replaced in a reliable fashion, they cannot safely be put in the service of creating embryos to secure elimination of disease in clinical settings (Newman, 2017a). Yet, bullish promotion of clinical use for genetically modified embryos is taking place at the historic moment that the underlying theoretical paradigm for successfully doing so is crumbling. Genes do not determine the features of organisms in the ways commonly believed and taught over the past century. As we will see, to compensate for the misalignment between moving forward with engineering embryos, on the one hand, and the unknowns cautioning against doing so, on the other, some advocates of the technologies seize upon troubling rhetorical strategies. These include false analogies, failure to account for new science relevant to technologies, ignoring troubling but predictable developments and, at times, outright deception. These methods effectively obscure crucial junctures and exit ramps on the already paved road to Gattaca.

Signposts Ignored: The Fallacy of Genetic Determinism

Profound problems with the very notion of "genetic engineering" cloud visibility on the road to Gattaca. Leaving aside compelling evidence that the accuracy of CRISPR has been overstated, even if its modifications could be made with perfect reliability, genes do not determine traits in any straightforward fashion. The scientific basis for this recognition was discussed in

Chapter 2, but the implications are vividly illustrated in behavioral studies in animal systems.

Mouse studies (the most popular surrogate, or "model," for analyzing the biology of humans) repeatedly disconfirm an iron law of genetic governance. Two investigations serve as examples. The goal of the first study, reported in 1999, was to compare outcomes of behavioral tests. The behaviors included locomotor activity in an open area and in a maze, with and without drug treatment, and learning to swim. Researchers conducted tests in three different laboratories (two in different regions of the U.S. and one in Canada). The investigators used exactly the same inbred strains and one mutant strain of mice and went to "extraordinary lengths" to equate test apparatus, testing protocols, and all aspects of the laboratory environments. They nonetheless found significant and, in some cases, large effects of the different sites for nearly all measured behaviors for each strain. Although the results showed that genetic differences between the strains had an impact on abilities to negotiate mazes, swim to a preselected goal, and so forth at each research site, the directions of the differences were in some cases reversed among sites (Crabbe et al., 1999).

The second study, from 2013, showed that the same gene can have different roles, not only between species (calling into question the validity of mice as human research models), but even within the same species. The study tested effects on gene activity or "expression" of the same kinds of inflammation-inducing stresses in mouse and human subjects. The investigators found that inflammatory stress resulting from burns, trauma, infection or sepsis led to very similar gene expression responses among the human subjects. In contrast, the genetic responses in mice under these conditions were not the same as the human ones, and in some cases, mice from different inbred strains responded differently from one another (Seok et al., 2013).

The take-home message from both these studies is that the roles of particular genes in mature organisms, even against a constant background of other genes (as in inbred mice) are unpredictable. For some traits in some species (e.g., response to inflammatory stress in humans), there may be evolved "homeostatic" or stabilizing mechanisms that enforce uniformity of gene activity, but scientists currently have very little idea of how this is achieved. What's true of mature organisms is all the more true of embryonic ones. Studies on animal embryos have shown that development can stay on track in the face of natural and experimental perturbations while the body is taking form. But the basis and limits of this channeling of developmental outcome (referred to as "canalization"), is even more obscure than the homeostasis of adult bodily functions.

For the human species, the scientific and social implications of such studies are far-reaching and powerful. Humans are much less genetically uniform than inbred strains of mice. It is most likely that the majority of attempts at genetically engineering them (that is, introducing genes during development that have not coevolved with either of the parents' genomes) will result in adverse outcomes. Committing to an enhancement strategy would mean pushing past these "bad results" and trying over again, hoping for an "improved" round of genetically engineered offspring. These hazards would apply equally to manipulations that most people might consider frivolous or cosmetic, such as eye color or athletic ability, as to those that would be less controversial, such as the elimination (in an affected embryo) of dread conditions such as Tay-Sachs or Huntington's diseases.

The inevitable experimental errors – children with profound disabilities from genetic engineering mishaps – would likely be viewed negatively by parents seeking biological improvement. As with any design-oriented technology, the quality control impulse sets in, motivating seekers of the unflawed to try again. What will be the moral and legal status of "experiments gone wrong"? Undeterred by the perplexity of such questions, aspiring human genetic engineers and their commercial backers continued to spin the feasibility of biologically "improving" the human race.

Misdirecting Signposts on the Road to Gattaca

Given its profound implications for the human species, proposed uses of CRISPR technology certainly warrant vigorous scrutiny (Jasanoff and Hurlbut, 2018). But decision-making about biotechnological development is the result of elite processes and market forces – and sometimes, deliberate misdirection. Whatever their usefulness in laboratory research on early embryogenesis may be, misleading characterizations about cloning, "three-parent" embryos, and human-animal chimeras, have encouraged discussion of clinical application of those technologies. To smooth their reception, proponents encouraged deliberate misunderstandings about them, making their use appear familiar and unworthy of concern. Their chronologies overlap, and normalization in one enables normalization of the others. Each time humanity eyed one of these biotechnologies it was standing at a crossroads. Each time, it could have stepped off the road to Gattaca – had the route underfoot not been determinedly disguised.

In the case of cloning, enthusiasts encouraged the mistaken belief that clones are twins (something familiar) although clones are not twins. This has encouraged the idea of using a highly experimental technology with demonstrated hazards to produce new humans by conflating it with something entirely natural. In the case of using genomes from three different people to create embryos, promoters encouraged calling the process "mitochondrial transfer." In fact, no **mitochondria** are transferred by the procedure. The misnomer served only to obscure the unprecedented mix of genetic material from three people to create embryos for implantation, a result made possible by building on cloning technology. Finally, advocates of mixing human and animal cells to create mostly non-human embryos assured the public that doing so would not bring us closer to creating mixed-species organisms that are "mostly human," or have human brains, even though the techniques enable doing just that (Regalado, 2016) – and some have proposed it (Begley, 2017b). These sleights of the rhetorical hand hid how policy recommendations for human biological modifications were gearing up to transgress boundaries. At times, achieving the illusions meant side-stepping accepted scientific knowledge, well described in popular textbooks and easily accessible research reports.

Cloning Confusions: Of Clones and Twins

The cloning of Dolly the sheep by Ian Wilmut and his associates led to a spate of proposals to clone humans. The justifications varied. Nathan Myhrvold (at the time Microsoft's chief technology officer), for example, declared that "[c]loning is the only predictable way to reproduce" (Myhrvold, 1997), and that resistance to fabricating people by new technologies was "racist." U.S. Senator Tom Harkin, addressing his colleagues, characterized opposition to human cloning as anti-science, asserting that "the attempt to limit human knowledge is demeaning." But most cloning enthusiasts attempted to normalize the effort by invoking false analogies between SCNT cloning and the familiar occurrence of twinning.

Twins (and triplets, quadruplets, etc.) can arise when the cells of an early embryo (blastomeres) separate from each other. If they are **monozygotic**, that is, if they arise from one fertilized egg, the resulting individuals are termed "identical." Even embryos that do not disaggregate in the normal course of gestation can be forced to do so. "Embryo splitting," resulting in two genetically identical embryos from one fertilized egg, has long been used in the cattle industry. Aldous Huxley in his 1931 novel *Brave New World* (Huxley, 1932) contemplated human applications of embryo splitting in the "Bokanovsky Process" in which up to 96 identical embryos were produced. (The actual limit in humans seems to be 16.) In Huxley's fictional world these were treated in specific ways to produce humans of different social categories.

The type of cloning used to create Dolly is called, "somatic cell nuclear transfer" or SCNT (see Chapter 2). To recall, in this process, the nucleus of one cell is removed and then placed in another cell that has had its nucleus removed. But clones generated in this way have little in common biologically with human twins or the cohorts from Huxley's imagined procedure. In the first place, genes in separated blastomeres of an embryonic twin function in egg cytoplasmic microenvironments of identical composition. The genes of a SCNT clone, however, operate in different egg cytoplasmic environments – and each additional SCNT clone of the same donor will develop in still different egg environments. It is well known that monozygotic siblings can be very different from one another in personality, abilities, health, and so forth despite having the same egg environment during early development. This is even truer for SCNT-produced animals, which are more different biologically from one another and from their genetic antecedent, the nucleus donor, than twins are from each other.

Second, natural identical twins do not have such a genetic antecedent. Monozygotics (even *Brave New World*'s) may have identical genes, but no other individuals have, or ever had, the same gene variants as they do, so they are all genetically *unprecedented*. However, the impulse to create SCNT mammals (like the cloning of a deceased pet) is tied to a desire for the new animal to be as much as possible like a previous one. While someone who undertakes this procedure may recognize that genetic replicas are not exact copies, the very act of producing them denies the new individual (the pet, or potentially a human) the biological uniqueness that pertained to each previous member of its species before SCNT became possible.

A third difference between monozygotics and SCNT clones – the most important one biologically – is that unlike the former, the latter are assembled from portions of severely damaged cells, i.e., an isolated nucleus and an egg that has had its own nucleus removed. It remains a scientific mystery as to how these cell fragments can recover from the insult and occasionally cooperate to produce an apparently fit member of the respective species. Evolution can only come up with means to protect against harmful variations (such damaged DNA or proteins) encountered commonly in the history of life. Because nuclei are never exchanged between the eggs of different individuals in the course of embryonic development this particular kind of biological derangement has never appeared pre-biotechnology.

The profound differences between SCNT clones and twins did not stop various influential commentators and scientific popularizers such as the Oxford professor Richard Dawkins (Dawkins, 1998) and the late Harvard professor Stephen Jay Gould (Gould, 1997, 1998) from justifying SCNT cloning by playing on fears that people might have that clones would be mere replicants rather than unique individuals. This very reasonable concern, given the radical manipulations involved in SCNT, was portrayed instead as scientific ignorance that must be squelched, a constant refrain aboard the biotech juggernaut.

Dawkins and Gould handled this dismissively by attempting to naturalize the prospect of SCNT cloning.

Gould wrote:

> Dolly the sheep, the first mammal cloned from the adult cell of a single parent [sic²] . . . has shocked the world beyond any merely intellectual reason – primarily by raising for so many people . . . our deepest worries about the distinctiveness of our own personhood . . . May I suggest . . . that all these fears are misplaced, for these questions have a clear answer, known to all human societies throughout history. Identical twins are clones . . . [y]et we know, and have always known, that human identical twins – whatever their quirky similarities in behavioral details, as well as physical appearance – become utterly distinct individuals.
>
> (Gould, 1997, pp. 15–16)

Gould wrote as if what was bothering critics about the prospect of producing SCNT humans was the fact that two humans would have the same genome, rather than the bizarre motives of those who might wish to produce a person from a *preexisting* prototype, or the biologically ill-advised enterprise of trying to make a human by putting pieces of cells together and hoping that nature will make it right (Gould, 1998). Richard Dawkins, for his part, suggested that anyone who makes a moral distinction between permitting the birth of naturally occurring twins, and forming a new individual from the genetic template of an existing one has a low IQ (Dawkins, 1998). Even Microsoft's Nathan Myhrvold chimed in with his own version of the deceptive characterization, declaring that "[c]lones already exist . . . We know them as identical twins" (Myhrvold, 1997).

Now that cloning methods have been sufficiently optimized to have normalized cloned farm animals, polo ponies, and cats and dogs (see Chapter 2), it is reasonable to ask whether society will soon be ready to accept full-term cloned humans. The first thing that must be recognized is that we don't know how cloning affects subtleties such as the wiring of the cerebral cortex, which contains about a half-billion neurons in a dog, and about 16 billion in a human. A cloned dog may be acceptable as a companion animal even if the development of its brain has been perturbed by the cloning process. Human insensitivity to and tolerance for variations in the mentality of nonhumans might make such errors acceptable. Tolerance of potentially altered minds of cloned humans, however, if such were produced, is less assured. Social experience indicates that they might eventually be accommodated by some fellow humans, but perhaps not by others.

Given these caveats, what might be the motivation to clone humans? There is no question that a clonal human would have a marked resemblance (albeit not "identity") to its genetic prototype. This might be sufficient to bring some solace to a person who has lost a child, sibling, or even a parent.

In a more practical sense, the advantages of cloning to overcome the uncertainties in the production of "savior siblings" has been discussed in Chapter 2. With the use of CRISPR-type gene manipulation, a sibling genetically congruent in all ways but the mutation can be produced for a child with a gene-related disease. But there would be health hazards in proceeding, and the precedent of refining and embellishing preexisting genetic prototypes would change human civilization in ways we can barely imagine.

Twins are humans, and so would clones be. But SCNT technology threatens to bring us to the point where human identity may become ambiguous. According to the tenets of genetic determinism, if a SCNT clone has a species' genome it is a member of that species. But CRISPR editing can be used to make any changes in human DNA a scientist may wish. If someone stitched some (or a lot of) mouse or worm DNA into a human nucleus before SCNT was performed what would be the status of the resulting organism?

To further elaborate on such scenarios, in what category should we place an animal developing from a sheep's egg containing a human nucleus (a current research model; Hosseini et al., 2015)? Other plausible, scientifically motivated SCNT manipulations include swapping out one or more of the 23 different human chromosomes for one from a nonhuman animal (Pereira et al., 2008), or substituting segments of human DNA with synthetic DNA sequences (Liskovykh et al., 2015). The way a SCNT clone is constructed enables such possibilities. Moreover, other than restrictions on federal funding, no federal laws in the United States restrict such efforts, even to the extent of bringing such engineered clones to term (FDA Guidance, 2015). The recent report from China that macaques (primates, like humans) have been successfully cloned (Cyranoski, 2018) makes speculations about producing humans via SCNT no longer academic.

Normalizing Three-Parent Children

In early 2015, the British Parliament approved clinical use of technologies that enable creating embryos possessing DNA material from three different individuals. In the US, the FDA discussed the procedures and although not approved by that agency, they are nonetheless being implemented by physicians (Reardon, 2016). Promoters touted the technologies as effective ways of "curing" mitochondrial diseases. Mitochondria are the cell **organelles** that extract energy from fuel molecules. These organelles have their own DNA that specify only 13 of the more than 800 proteins that function within them. (There are 24 other mitochondrial genes that specify RNA molecules.) Mutations in genes that specify mitochondrial proteins, which affect about 1 in 4000 births, can adversely affect hearing, vision, pancreatic function, and neuromuscular activity, among other physiological systems, but most of these mutations are in nuclear genes, located outside the mitochondria. New techniques were approved to create infants who would inherit only nuclear genes from women whose mitochondrial genes carry impairing mutations. These infants would obtain the needed mitochondrial genes from an egg donor. This is not a "cure," in the usual sense of the term, and also is ineffective against mitochondrial disease whose origin is in nuclear genes.

The two approved methods involve inserting a cell nucleus isolated from the egg of one woman into an egg of another woman (the nucleus of which is removed). This can be done either before or after fertilization. When done beforehand, it is called "**maternal spindle transfer**" (**MST**). When it is done after fertilization it is called, "**pronuclear transfer**" (**PNT**) (Wolf et al., 2015). When fertilization does take place, genetic material from three persons would then have formed the embryo ready for implantation into a uterus (possibly of a fourth person, a woman different from the two cell-fragment donors). All throughout the period of deliberation around their approval these techniques were referred to as mitochondrial "transfer" or "replacement" by their scientist-creators, journalists, bioethicists, members of regulatory panels, and legislators (Chinnery et al., 2014; Ishii, 2014; Ridley, 2015; Wolf et al., 2015). But, since no transfer of mitochondria is involved in the procedures, these descriptions are scientifically inaccurate (Newman, 2014). In her zeal to quell opposition to its implementation, bioethicist Nita Farahany, a member of President Obama's Bioethics Commission, described it in the *Washington Post* as exactly the opposite of what it is: "Put simply, the mitochondria of the affected egg are removed and replaced by the mitochondria from the healthy egg" (Farahany, 2014).

In fact, both MST and PNT are similar to cloning by somatic cell nuclear transfer (SCNT). Like SCNT cloning, the techniques involve replacement of an egg's nucleus by a nucleus from another woman's egg. From the viewpoint of the woman with impaired mitochondria, her nuclear genes will eventually coexist in her prospective child with the mitochondria of the egg from the second woman which, being different from her own, she might consider, "replacements." But the mitochondria are "replaced" only in the same way that someone might misleadingly describe their windows as being "replaced" when what they really have done is to move into a new house. This is so because in addition to mitochondria, the second woman also provides all the non-nuclear components of her egg (the **cytoplasm** and membrane, with their hundreds of subcomponents), which are well known to play major roles in determining the characteristics of the offspring in mammals (Kloc et al., 2012).

The fact that the egg that will be implanted in someone's uterus originated in the body of the second woman means that if *she* had legally contracted to be the baby's mother the manipulation would have been characterized as a "genome transfer" or "genome replacement." The most neutral characterization of the procedure is that an infant is produced using an egg that has had its nucleus (including its nuclear genes) removed and replaced by a nucleus obtained from a different person's egg.

Because a different woman from the egg producer provides an egg nucleus with its 20,000 plus genes, standard genetic determinism makes it natural for advocates of the "three-parent" (or three-person, "parent" used here only in the strict sense of biological progenitor), procedures, whether researchers, physicians, or bioethicists and science writers to invariably refer to the nucleus donor as the "mother." In fact, the woman who provides the egg, or who gestates the embryo, also has biological and social claims on being the mother, but only the paying client is referred to in this fashion. Supporters also misrepresented the nuclear transfer procedure as the replacement of a few genes (the mitochondrion has just 37 of them) and thus characterized the technique as a trivial manipulation. Actually the full 20,000 genes are being replaced, if we want to use a gene-centric framing. Uncritically parroting the mischaracterization, the journal *Nature* reported that the technique's developers "have compared mitochondrial replacement to changing the batteries in a camera" (Callaway, 2014).

Why does it matter whether the procedure is represented as the replacement of a few dozen genes or thousands? It matters because animal studies during the deliberations were showing the technology to be unsafe, leading to anomalies such as altered susceptibility to diabetes and to cardiac damage due to incompatibilities between nuclei and mitochondria that have not coevolved (Gershoni et al., 2014; Dunham-Snary and Ballinger, 2015). Though there were hints of this earlier, the issue was not discussed to any significant extent by critics during the approval process, nor was it brought up at all by scientific advocates or their bioethicist allies.

The animal studies presented potential difficulties for public perception and threatened shepherding it easily through the regulatory process, including hearings before Parliament. The misleading framing of the procedure ("only 37 genes"; "replacing batteries") was highly effective in surmounting any opposition to certifying the three-person procedure for clinical use in the U.K. Pitching it in terms of "20,000 genes" or "cloning" would not have played nearly as well. As of this writing (March 2018) two women in the U.K. who have symptoms related to impaired mitochondria, have been approved to have their next babies by the three-person procedure (BBC News, 2018). Whatever experimental errors might be introduced in these offspring by these uncertain procedures will be passed on to their own descendants.

Finally, it is an irony of the caution-laden (though, as we have seen prone to deception) deliberative process around three-person babies in the U.K. that under current circumstances rogue medical scientists can readily take things into their own hands and sell the technology to willing customers with impunity (Cha, 2018). The international entanglements can be quite baroque. In a widely discussed case, for example, Dr. John Zhang, mentioned above, navigated through and possibly violated the regulations of several countries by performing MST in Mexico in 2015 for a Jordanian couple who were patients of his U.S. clinic. The prospective mother had impaired mitochondria. A child claimed to be healthy (though not examined by independent physicians) was born from the three-person embryo, though with a mixture of normal and abnormal mitochondria that is often concerning with respect to future health (Zhang et al., 2017). Purveyors of these procedures seem unconcerned with the fact that they are doing uncontrolled experiments on prospective human beings, however. As the proprietor of the Ukrainian clinic using three-person procedures for infertility of unknown etiology (described above) stated to a reporter "The only way to know whether the procedure works and is safe is to try it . . .. If you would like to swim, then, first of all, you must jump in the water" (Stein, 2018).

Human–Nonhuman Chimeras

A clear threat to human biological integrity would be the creation of embryos that are part human and part animal. Chimeras are embryos and full term animals created from embryonic cell mixtures from differing animals. (The term was originally used by the ancient Greeks to refer to monstrous creatures containing parts and characteristics of multiple animals.) Beginning with mice, rats, and rabbits in the early 1970s, researchers succeeded in 1984 with the dramatic creation of full term goat–sheep chimeras, whimsically dubbed "geeps."

Chimeras are not members of a single species. As organisms that develop from mixtures of embryonic cells of two different species of vertebrate animals, they are unprecedented in nature and are thus in an important sense human-made "artifacts." Biomedical researchers, commercializers, and regulators have stepped forward eagerly with reassurances that this technology can be kept under control (Stein, 2016). In fact, it is susceptible to ready misuse. Although there are now ways of targeting the human cellular contribution to particular organs, there is no reliable way to prevent someone from using the same methods to produce organisms that have a high proportion of human cells across all organs, including the brain. And the motivations to do so are great: there are many potential scientific and medical uses for animals with increasingly larger human contributions, particularly regarding studies of different organ systems and their responses to stress, experimental drugs, etc., as well as prospective advocates for patients needing tissue replacements (Newman, 2006, 2017b). The promoters of these procedures have thus helped to mainstream a research practice once dismissed by the scientific community as outrageous.

This technology first came to public attention in 1984 with reports of the goat–sheep chimeras – "geeps" (Meinecke-Tillmann and Meinecke, 1984; Fehilly et al., 1984). Embryo chimeras are different from animal **hybrids** produced by cross-fertilization. Mules (donkey–horse hybrids) are the most familiar examples of the latter organisms. In a chimera, the cells used to construct the new animal retain their species identity, unlike a hybrid, where each of the cells has a mixed identity. While chimeras and hybrids are both intermediate in biological character between the two originating species (a geep does not look exactly like a sheep or a goat, but looks somewhat like each), the fact that a chimera is a mosaic rather than a blend of two species at the level of its cells has bizarre and unsettling implications. Each egg or sperm cell in the ovaries or testes of a female or male chimera arises from a precursor cell contributed to the embryo by one or the other species. They are therefore genuine goat or sheep eggs or sperm. If two geeps mate, the outcome would never be another geep, because geeps are constructed by mixing embryo cells, not by fertilization. Successful mating of geeps would have three different possible outcomes: (i) a goat–sheep hybrid (if a goat sperm or egg met up with a sheep sperm or egg), (ii) a goat (if a goat sperm fertilized a goat egg), or a sheep (if a sheep sperm fertilized a sheep egg). Correspondingly, two pig–human chimeras could potentially birth a human.

The successful demonstration of the chimerization technique raised questions about human applications, and part-humans as "inventions." Since the 1980 *Chakrabarty* decision approving patents on living organisms and their descendants, Congress had drawn no clear line precluding the patenting of an appropriately modified human embryo. Neither had it clarified the limits of patentability of human–animal chimeras: how many human genes or cells would a non-human animal have to possess before it could *not* be patented owing to Constitutional protections of members of the human community? (Parthasarathy, 2017).

In 1997, *Biotech Juggernaut* co-author and developmental biologist, Stuart Newman, decided to test boundaries and tease out some of the legal and ethical ramifications of conducting scientific research in the post *Chakrabarty*–Bayh-Dole world. Newman, assisted by Jeremy Rifkin, President of the Washington D.C. based, Foundation on Economic Trends, wanted to do so in a fashion dramatic enough to attract public attention. Never actually intending to create them, in fact, hoping to discourage others who might plan to do so, he applied for a patent on chimeric embryos and animals that would contain both human and non-human cells.

Newman's application included proposals for combinations of human and mouse cells, as well as human and chimpanzee cells, and could contain anywhere from a minor proportion to a majority of human cells. In addition to describing how such chimera would prove useful for study by developmental biologists, the application suggested that partly human embryos could serve as sources of transplantable tissues and organs for human patients and would be useful for testing the toxicity of drugs and chemicals.

Newman's goal was to raise questions about the difficulties of drawing consistent and defensible lines between what is a human being and what is not, and what is a life-form and what is an invention. A related objective was to alert the broader public about where these new technologies could lead (Dowie, 2004; Newman, 1997, 2006). While most people may not have been bothered by a human with a few animal cells, or an animal with a few human cells, there would be some degrees of mixture that would likely disturb anyone. Because the embodiments of the invention were to be chimeras – animals made by mixing cells from early embryos of two different species – they had a "composite identity" from their inception. This is very different from an organism such as a human with a transplanted pig valve, whose species identity is unambiguously established during development and only receives the foreign tissue contribution afterwards. Patents can only be awarded for useful inventions (even if the item is never to be built). As noted, descriptions of utility for the chimera patent included uses of partly human embryos and full-term animals in developmental biological research, in toxicology, drug testing and in regenerative and reparative medicine.

Although the motivation was publicly acknowledged, the stratagem triggered numerous law review articles dealing with the Constitutional issues it raised (e.g., (Magnani, 1999; Coughlin, 2006; Heled, 2014)) and caused significant alarm in the legal and scientific communities. The chair of the department of genetics at Harvard Medical School, the scientist who patented the first mammal (a tumor-prone mouse used for cancer research), told an interviewer, "[t]he creation of chimeras is an outlandish undertaking. No one is trying to do it at present, certainly not involving human beings" (Zwerdling, 1998). In 1999 the U.S. Patent and Trademark Office (PTO) rejected the invention on the grounds that some of its embodiments would "embrace a human being" (Weiss, 1999), and held to this position through a series of amendments responding to arguments for rejection by patent examiners, until the final submission in 2005 (Weiss, 2005; Newman, 2006). The last decision also was a rejection, but Newman decided not to contest it, judging that his main objective, alerting the public, had been accomplished (Newman, 2006). But the alert, though prescient, would not stem the tide. While Newman's patent application worked its way through the legal system with the hope that troubling questions at the intersection of biology and culture would receive maximum public deliberation, chimeric research agenda stealthily navigated the controversy.

In 2002, a Rockefeller University scientist proposed injecting human embryo stem (ES) cells into mouse embryos. The purpose, he said, was to investigate the development and therapeutic potential of the ES cells. Much like the Newman patent application, this genuine experimental proposal drew resistance. In November, the New York Academy of Sciences held a meeting to deal with the opposition brewing on the part of some developmental biologists against the proposal. One of the attendees, Ronald McKay, a leading stem cell researcher at the National Institute of Neurological Disorders and Stroke (NINDS), voiced disapproval: "I am completely opposed to putting human embryonic stem cells into any condition that will cause moral affront," he stated. Others who were critical suggested ways to accomplish the same research goals without creating such human-animal chimeras (DeWitt, 2002). While controversy brewed, proponents of the chimera agenda grew secretive. Investigators wanting to pursue the chimera protocol held a closed session, excluding other participants from the New York Academy meeting. One of those barred from attendance was James Battey, another researcher from the NINDS, who was also chair of the National Institutes of Health Stem Cell Task Force. He criticized the "excessive secrecy" of the advocates (DeWitt, 2002).

After the 2002 meeting, chimeric research continued largely without public attention. Regulations enacted in tandem with the research pertained only to federal funding. This source of support is an important stimulus to this kind of work, but as we have seen with California's and other states' willingness to step in with their residents' taxes to make up the deficit (see Chapters 3 and 4), its absence is not a deal-breaker. By 2006 a group at Rockefeller University published a paper titled, "Contribution of Human Embryonic Stem Cells to Mouse Blastocysts" (James et al., 2006). In 2007, a University of Nevada scientist created a sheep after injecting human adult stem cells into a sheep's fetus (Almeida-Porada, 2007). In 2009, the National Institutes of Health (NIH) promulgated funding guidelines for human-animal chimeric research (NIH Guidelines, 2009). The guidelines prohibited funding for the breeding of chimeric animals in which human stem cells may have become eggs or sperm, to preclude the "human-babies-from-nonhuman-parents" scenario described above. The NIH also declined to fund experiments involving chimeras between humans and nonhuman primates in advisories published in 2009 and 2015.

Funding restrictions are not bans, however, and some of the proscribed research nonetheless went forward with non-NIH support. Research funded by the state-sponsored California Institute of Regenerative Medicine (CIRM) (see Chapters 3 and 4)[3] advanced this effort with the help of what are called iPS cells. CIRM was created in 2004 in large part to support research with human embryo stem (ES) cells (Longaker et al., 2007; Trounson et al., 2008). Through the Nobel Prize-winning efforts of Shinya Yamanaka and his colleagues, however, it became possible to use substitute cells in place of ES cells – induced pluripotent stem cells (iPSC). These cells, which function like ES cells, are produced by activating a small number of developmentally early-acting genes in somatic (non-reproductive) cells of the mature body (Yamanaka and Takahashi, 2006). While iPS cells cannot form human embryos on their own, they can generate any portion or percentage of a chimera (the human component) if inserted into an actual embryo, either human or animal.[4]

What this means is that while the *origin* of iPS cells is not in a human embryo (which is what made ES cells a source of controversy), their *fate* in chimeras is to generate portions, which can be arbitrarily large, of a human body or brain. This is a disturbing prospect. While the chimeric animals currently being produced are "mostly non-human," as we have noted, the exact same techniques could lead to chimeric organisms that are "mostly human."

Members of the stem cell research community launched a concerted lobbying campaign (Sharma et al., 2015). "Tremendous potential exists to develop humanized disease models for studying drug pharmacology," they claimed, "illuminate genetic disease pathogeneses . . . [and] to generate an unlimited supply of therapeutic replacement organs." There was little concrete evidence to support the assertions. Even so, they were effective in finally cajoling the NIH to throw in the towel. In the summer of 2016, the NIH issued an advisory lifting the ban on funding human–animal chimera research (Kaiser, 2016). In fewer than 20 years and without public debate, research once considered "outlandish" and beyond the pale by prestigious members of the scientific community now qualified for public funding.

Some business models, fear or vanity based, make consumer demand for violations of human biological integrity plausible. A 2017 *New Yorker* article on celebrity entrepreneurs touting genetic engineering provides some vivid examples. These include proposals to use longevity genes identified in dogs and transfusions of young people's blood to extend their customers' lifespans (Friend, 2017). A commercial startup, Nectome, received a contract from MIT to develop methods for physician-assisted suicide that preserved the patient's brain for eventual placement in a healthy host (perhaps a body donor with an accidentally damaged or cancer-ridden brain; Regalado, 2018). Though the contract was cancelled, Nectome continues to pursue this project. Although this company's service does not employ human–animal chimeras, its bizarre character, now a commonplace on the biotechnology scene, suggests extensions that do. For example, it is easy to imagine a company eventually marketing the mixing of a farm animal embryo (e.g., sheep or pig) with a patient's brain stem cells, with the grown chimera providing an intermediate host for a brain genetically equivalent to the patient's.

Certainly, there is no law or regulation to curtail it. And the covert way that initially exotic projects have of finding acceptance by the bioscientific establishment thwarts the influence of countervailing cultural norms.

Japan, long a bastion of restraint in human applications of biotechnology, provides a striking example of the bumpy ride on the road to Gattaca. In 2000 the Japanese government made human cloning for reproductive purposes a crime, for example. But a government-appointed expert panel has nonetheless recently (in mid-2018) advised lifting a long-standing ban on creating human–animal chimeras for human organ cultivation and harvesting (see "Japan to Lift Ban," 2018).

"Solutions" to previously intractable problems, discomfiting when first mentioned, have ways of moving onto the medical agenda. From there, as we have seen, they become normalized as part of the professional mission of adherents and their bioethicist associates. Thirty years ago the notion of three-parent children and cloned human embryos would have been met with wide revulsion, as would human–animal chimeras of any kind. Now these entities are just part of the research and clinical landscapes.

* * * * * * * * *

Not so long ago, making embryos and full-term humans from bits and pieces of several people's eggs and somatic tissues, and creating embryos from human and animal cells would have been seen as troubling experiments in human development. But in the blink of an historic eye, human embryo cloning, construction of three-parent babies, and human–animal chimera have been implemented or discussed in favorable ways. Their professional normalization raises red flags signaling the role of unforthcoming or outright deceptive strategies in ushering into public acceptance experiments that threaten human biological integrity and cultural norms. Proponents apparently perceived a need for deceptive strategies to overcome public resistance.

Along the way, unsupportable claims were made: that human clones are the same as identical twins, that egg nucleus replacement was actually merely mitochondrial replacement, and that only 37 genes are transferred in the three-parent procedure instead of 20,000. Erroneous claims were rarely disputed even by scientifically knowledgeable proponents. Some scientists even asserted (as we saw in Chapter 3) that clonal embryos are not really embryos. In the case of chimeric embryos, the potential for misuse was left unaddressed as research proceeded. As of now, the judgment of whether or not a chimeric entity displays "too human" a mentality to proceed with experimentally is left solely to the detection and discretion of the researcher (Lavazza and Massimini, 2018). The portrayal of these biologically profound manipulations as innocuous enabled not only their acceptance but also the "mission creep" referenced above: immediately after governmental approval of the three-parent techniques in the U.K., its developer started a campaign, and a company, to promote the use of the nuclear transfer technique for any woman who might fear age-related infertility, not just for disease prevention (Connor, 2015).

Avoiding Gattaca?

The impulse to genetically alter offspring arises from the desire to make them biologically "better" than they otherwise would have been. Our ability to predict how an introduced gene will behave in a novel context, however, (even a usually "normal" gene) is imperfect. Occasionally, genetic manipulations will lead to unexpected congenital anomalies – new forms of blindness, deafness, limb reduction, cognitive impairment – not all initially detectable. Inevitably, the drive to improvement will engender heightened discontent when inherently fallible methods do not work out as intended: unsuccessful products, though human beings, will be devalued. On the other hand, since some attempts will undoubtedly appear to be improvements, we might anticipate the rise of a pernicious consumerist ethos for human production, what has been termed "Yuppie Eugenics" (Hubbard and Newman, 2002; see also King, 2017).

The entry of different aspects and versions of the human organism into the marketplace represents a sea change in our civilization. As we have seen, failure to account for new science, failure to acknowledge the potential for misuse, false analogies, and, occasionally, outright deception worked to hide radical transformation. In other cases, where analysts have made a serious attempt to deal with philosophical issues raised by the construction of quasi-human animals, they have focused on questions such as the moral status of the interspecies beings (Robert and Baylis, 2003). But broader impacts on a society or culture in which the "human" has become a fluid or fungible quantity, as portrayed in *Gattaca* or Aldous Huxley's *Brave New World* before it, are, for the most part, ignored. David King, molecular biologist and founder of the British public interest organization, Human Genetics Alert characterizes the shifting terrain with sobering perception. "Once we begin to consciously design ourselves," he cautions, "we will have entered a completely new era of human history, in which human subjects, rather than being accepted as they are will become just another kind of object, shaped according to parental whims and market forces" (King, n.d.).

One aspect of the new reproductive technologies that will almost inevitably draw us down unanticipated and treacherous paths is that not all the products of the manipulations described are intended to be brought to full term. Some, like clones of patients, might be used simply to harvest replacement tissues, for example. But the same methods readily can be repurposed for full-term cloning. Notwithstanding good intentions, the lines between organ farming, human consciousness in animal bodies, and experimentally harmed children are permeable. And there will always be over-zealous medical innovators and venture capitalists ready to push the envelope and willing to have others face the risks.

Experience teaches that people accommodate to unfamiliar developments in surprisingly short periods of time, one development sliding furtively into another. As we have seen with human–animal chimeras, science fiction scenarios or "outlandish" provocations by biotechnology critics in one period can become ordinary science practice a few years later.

How science is practiced and how scientific research is funded makes even research that is clearly circumscribed, such as that of Kathy Niakan, susceptible to advancing a eugenic future. In September 2017, *Nature* published a report summarizing research undertaken by her team at London's Francis Crick Institute. Her study used CRISPR-Cas9 editing to investigate human embryogenesis (Fogarty et al., 2017). Niakan sought to examine the mechanisms of early human development to explore the basis of infertility and premature cessation of pregnancy. While her work was enabling the extension of the time human embryos could be gestated outside of the body (an objective meeting with ever-increasing success; Shen, 2018; see also Khazan, 2017), Niakan stated that she did not want to produce full-term, genetically modified people.

Notwithstanding her own scientific objectives, however, nothing prevents other scientists from using Niakan's findings as a stepping stone to what, for some, is the ultimate goal of fabricating full-term humans. This is so because the practice of science is a collective activity whose intellectual and tangible products always are subject to forces beyond the objectives of an individual investigator (Merton, 1968). Eventually, sufficient momentum gathers to cross borders once found too distressing to breach, as we have seen. Regarding "intended humans" (embryos implanted for reproduction), the only real way to avoid problems arising from technological mishaps is to not modify human embryos at all (see discussion in Newman, 2017b).[5]

Because it would involve experimentation with irreversible consequences, based on incomplete knowledge of genes (with the very concept of the gene being questioned), germline alteration is the "red line" that many have felt should not be crossed. The term "germline" emphasizes the fact that engineering embryos would typically cause transmission of genetic alterations to all future offspring of the engineered person. Although germline alteration dominates discussions about human genetic modification, this aspect would not be the most profound cultural watershed represented by the procedure. Once we accept the engineering of prospective people, even without germline propagation, something unprecedented in the history of human civilization occurs: the trial-and-error-based quality control ethos of industrial manufacturing becomes part of human procreation.

Even the most precise alteration of a known gene with CRISPR is fraught with uncertainties, including the possibility of inducing cancer (Begley, 2018). This may be worth the risk in an existing person with a disabling or mortal condition for which there is no other effective treatment. But it could never be so in an embryo, where the intention would be to improve a prospective individual's biological characteristics. Certainly a trait could be altered by gene editing, but not without the uncontrollable possibility of deranging other traits that may well have turned out normally in the unmodified embryo. Stated differently, organisms are not mechanisms or machines, and cannot be engineered as if they were (Newman, 2017a).

Doubtless, further research will make it possible to change a gene in all the cells of a prospective person, except for the germline. Then, potentially, parents can have a child with the enhancements they want without fear of transmitting errors to future generations. This technique will be presented as a responsible way of seeing if the methods will work. But it will be a technique that allows for all the social divisiveness associated with human perfectionism that reasonable people, skeptical of technological and commercial overreach, wish to avoid.

Successful work by researchers who find ways to study the early human embryo in vitro can be used as a platform to extend the number of days that embryos may be experimented upon before being destroyed. Regardless of their own wishes in the matter, these scientists have little say over where their work will be taken. In this respect, science has its own momentum. Existing regulations permit experimentation on a human embryo only before it develops its "primitive streak," an early indication of the developing central nervous system that emerges at 14 days of development. At a 2016 meeting of the California Institute of Regenerative Medicine (CIRM), however, bioentrepreneur and Nobel Laureate David Baltimore not only advocated for CIRM's funding of gene editing embryos he also urged increasing the number of days allowed for keeping embryos alive: "why 12 days and not 14 days? . . . everybody else talks about 14. Just want to give people the most freedom . . . so I want to think about when you won't want to prohibit that" (CIRM Notes, 2016, 127–128). When University of Cambridge human embryologist Magdelena Zernicka-Goetz and her colleagues demonstrated that it was possible to grow in vitro produced human embryos outside the body for 14 days (the previous limit was a week, Shahbazi et al., 2016), the Case Western Reserve bioethicist Insoo Hyun wrote "Now there will be further questions about whether or not there would be good scientific reasons for moving that line [i.e., the 14-day legal limit] out a little bit farther" (Hyun, 2016). There are strong indications that Baltimore favors production of genetically modified full-term infants (Pei et al., 2017).[6] But whether or not Baltimore or Hyun would wish their proposals to culminate in that eventuality, other powerful stimuli work in that direction. One such impetus is the federal research funding imperative to seek increased human applications. Another is the commercial interests and patent holdings sought by most of the investigators involved (George Church and Shoukrat Mitapilov being only two prominent examples[7]). Both forces pour grease continuously on the already slippery slope.

Given the influence of pro-technology interests, it is unlikely that genetic modification of embryos, already initiated, can be brought to a complete halt. No university-based reproductive bioethicists have publicly expressed doubt about the ability of the program to deliver on its promises. Mainstream bioethicists appear largely unwilling or unable to recognize challenges to received concepts in biology. Often their careers are tagged to benign consideration of the biotechnologies and science culture they evaluate (Elliott, 2001; Stevens, 2014.) Some may issue warnings about the negative potential of modifying prospective humans, focusing on increasing inequality, or devaluing those congenitally impaired (Singer, 2003; Evans and Moreno, 2015). But they almost never challenge the capability of scientists to make the improvements they propose. Genetic determinism, however erroneous, is such a compelling ideology, and the reality of developmental processes so complex and multi-causal, that there is little chance that analysts and regulators will appreciate the recklessness of the effort before it is well underway.

Then there is the prospect that the "bad" outcomes will actually find uses – in research laboratories where they will be used to study developmental processes, in organ donation clinics, and in pharmaceutical companies seeking humanoid experimental models on which to test new drugs. Once we begin thinking of human-type organisms not as anybody's children or parents (as opposed to mere donors of cells or genes), a whole new world opens up for biotechnology to exploit. In the words of the Stanford University professor and bioentrepreneur Drew Endy,

> If you look at human beings as we are today, one would have to ask how much of our own design is constrained by the fact that we have to be able to reproduce . . . If you could complement evolution with a secondary path, decode a genome, take it off-line to the level of information . . . we can then design whatever we want, and recompile it . . . At that point, you can make disposable biological systems that don't have to produce offspring.
>
> (Specter, 2009)

Gattaca explores social ramifications of what it means for civilization to embrace a society in full acceptance of eugenics – a term meaning "better inheritance." Left unaddressed by the film is how techno-eugenic societies will avoid creation of entities that are not inheritors of "better" traits but, rather, traits that are less desirable in the eyes of the designers or their clients. Huxley's *Brave New World* anticipated this: humans deliberately made to have low intelligence, the Gammas and Deltas, were assigned to subordinate roles in the social hierarchy. This inverts classic eugenic strategies – better inheritance gives way to worse – and we can speak then of "**meiogenics**" (Newman, 2012) from the Greek word for "less."

In our contemporary society, under the impetus of the biotech juggernaut, meiogenics takes on new implications as many barriers fall aside. Few may be contemplating producing lower orders of humans (though Endy's pronouncements give pause), but the demands of scientific experimentation and regenerative medicine for suitable biological materials are voracious, as we have seen in the case of human eggs (see Chapter 4). For attempts at human embryo engineering gone sufficiently "wrong" (e.g., producing brain-impaired fetuses), the outcomes might be salvageable for research or spare parts.

At present, existing regulatory regimes on human experimentation pertain to what are agreed-upon humans. Other, more permissive experimental regimes, cover vertebrate animals. In the future, synthetic biologists may be able to calibrate and titrate biological humanity and its animal consciousness. They would do this by taking the human genome offline and recompiling it. If so, we may be faced in a few decades time with unappreciated realities: all manner of humanoid organisms and animal–human combinations. Stanford's Endy has only become more avid in the past decade in envisioning a future of genetically engineered humans. He has even devised a new value system to accompany his program: a crowd-sourced "sense of the sacred in the realm of the digital." He amplified this in a recent interview, suggesting that in "20, 30 years from now ... [when] we're making some changes to a human genome, and it's actually one that we're going to bring to life," the effort can be imbued with sufficient gravity and social meaning through "a Wikipedia-type effort around writing genomes" (Duncan, 2018).

This will presumably become all-the-easier when emerging technologies to build "synthetic embryos" are perfected (Roberts, 2018). Here fertilization is performed using an egg and a sperm that are each generated by iPS cells, which by definition are abundant and replenishable. The iPS cells used in the procedure can also be genetically modified to combine and obscure relationships to any unique human antecedents. With embryos finally becoming expendable reagents, Endy's and other speculative bioentrepreneurs' Gattaca-like fantasies might actually become realizable.

Unwarranted confidence that embryo research can be halted short of full-term experimental humans, of either the eugenic or meiogenic varieties, undergirds trends now normalizing progression toward Gattaca. Without reversing this trajectory, it is difficult to see how we will not ultimately be led to socially divisive calls for new boundaries among various categories of hits and misses, some, as we have seen, relegated to merely practical uses in research or medicine (Newman, 2003).

Coda

Since germline genetic engineering of embryos is underway without the once sought after "social consensus," what is the social function of the public or quasi-public meetings that consider whether or not to move forward with these technologies? Has *Nature's* editorial caution that scientists not engage public discussion of the film *Gattaca* prevailed?

A meeting in February 2016 at the California Institute of Regenerative Medicine (CIRM) is instructive. Attendees considered whether the institute was empowered to fund research that involves the genetic "editing" of human embryos (Chapter 4). The presence of guest presenter, Nobel Laureate and bioentrepreneur, David Baltimore, harkened back to the 1975 Asilomar conference in Pacific Grove, California. There, Baltimore helped secure the future of "genetic engineering" when scientists gathered to discuss some of the social consequences of recombinant DNA research. Then, as now, conveners held at bay broad consideration of social justice and inclusive civic participation.

Baltimore spoke assuredly at CIRM's meeting. He saw no reason not to bring gene editing of human embryos into clinical use once it is developed. In the eventual ability to genetically enhance children, he expressed complete confidence:

> [T]he technology ... is not perfected today ... [Y]ou can make the simple argument we shouldn't do this because it's dangerous because we just don't know how to handle these powerful techniques, and that's true, but ... we'll solve those problems. They're too easily defined not to be solved.
>
> (CIRM, 2016, p. 26)

When asked whether safety was the only concern or whether we should also consider whether altering "mother nature" was advisable, Baltimore demurred that such "feelings" were "hard to evaluate" (p. 45). But the *feeling* that all problems using CRISPR to genetically manipulate embryos will be solved is not a matter of fact. It is a matter of faith. The counterindications are many and daunting – problems that promoters ignore.

When Dorothy Roberts, a professor of law and sociology at the University of Pennsylvania, asked whether social justice and disability rights communities shouldn't also be included in "discussions about what's ethical and appropriate for continuing down this path," an empaneled bioethicist replied that these communities need to be part of the conversation but added that they (CIRM) are limited "on the amount of time we have to actually have formal panels." She hoped that these communities would instead take, "advantage of the public comment period" and "send in materials." CIRM, she added, could also contract for white papers to cover "things that are not part of the panels that we're putting together" (p. 80).

Marcy Darnovsky, Director of Center for Genetics and Society, challenged a presenter for "graying out" an important set of social justice concerns. CIRM, she counseled, as a publicly funded program, needed to consider how to benefit

> the greatest number of people with the use of public funds . . . . [and] look at the kind of social dynamics and commercial dynamics that would be set in motion if germline modification were to be approved and put into a marketing context of a commercial fertility industry.

The speaker asserted that he had been asked "to present a fairly narrow, frankly just a few paragraphs in the entire document, but certainly justice is one of those principles that would fall out" (CIRM, 2016, p. 107). No time for justice. Earlier, David Baltimore, who chaired the 12 person committee that organized the 2015 international "gene summit" in Washington DC shared something of how organizers of that meeting had managed their time constraints: "We issued a final statement. That statement was issued by the organizing committee in its own name and is not an official document of the academies . . . We actually wrote a lot of it ahead of time, full disclosure," he confided, "because otherwise we wouldn't have been able to do it. And there are four major conclusions." The first conclusion was, "both basic and preclinical research should continue on the editing technologies and on the benefits that can result from editing and on the biology of human embryos and germline cells" (CIRM, 2016, p. 29). Having arrived with this recommendation in hand, could rigorous debate really have been expected?

Public meetings organized by bioscientists and bioethicist colleagues find no room for serious, even-handed consideration of broadly based social reasons for halting genetic modification of human embryos. More unlikely still is that attendees would ever agree to create regulations designed to prevent privately financed human GMO projects or curtail those currently underway. Even should critics be invited to the table, it is highly doubtful that meetings will ever be more than procedural strategy for avoiding substantive change of direction. Short of sustained challenge from an informed, exercised citizenry, there is scant hope of falling back from the steady march to Gattaca.

Notes

1 Portions of Chapter 6 are from Newman, S.A. (2017b), "Sex, Lies, and Genetic Engineering: Why We Must (But Won't) Ban Human Embryo Modification," I. Braverman (ed.), *Gene Editing, Law, and the Environment: Life Beyond the Human* (Abingdon, U.K.: Routledge), 133–151.
2 The donor of the mammary gland cell nucleus used to produce Dolly was in no fashion her "parent," either biologically, or in the nurturant-maternal sense. Like the identification of SCNT with twinning this erroneous description conflates an artificial manipulation with a familiar one in the service of naturalizing it.
3 See www.cirm.ca.gov/about-cirm/our-mission
4 The availability of iPS cells has also mooted proposals to use molecular biological means to circumvent the use of embryos to obtain ES cells. (For such a proposal, see Hurlbut, 2005; for the suggestion that ES cells were no longer needed, see Byrnes, 2005.)
5 A proposed ban on modifying human embryos has no bearing, of course, on a woman's right to terminate a pregnancy.
6 This paper, of which Baltimore is a co-author, contains many caveats regarding safety and general societal acceptance of clinical, i.e., full-term applications of human embryo gene editing. But it also contains the following passage:

> Recently, Ma and colleagues reported the correction of a four-base GAGT deletion in MYBPC3, one of the genes implicated in dominant hypertrophic cardiomyopathy, in pre-implantation human embryos (Ma et al., 2017). In contrast to earlier reports, the new study not only used diploid human embryos but also claimed improved efficiency, accuracy, and reduced mosaicism, thus breaking new ground toward clinical use of this approach.

7 Church has co-founded "scores" of companies, including Ediatas Medicine, Veritas Genetics, and GRO Biosciences, www.bizjournals.com/boston/news/2017/09/28/george-church-startup-bags-2m-in-quest-to-create.html. Mitalipov founded Mitogenome Therapeutics and is involved in commercial ventures on genome engineering with Korean and Chinese companies, www.biosciencetechnology.com/article/2015/03/us-cloning-pioneers-startling-new-partnerships

Sources Consulted for Chapter 6

Almeida-Porada, G., C. Porada, N. Gupta, A. Torabi, D. Thain, and E.D. Zanjani, "The Human–Sheep Chimeras as a Model for Human Stem Cell Mobilization and Evaluation of Hematopoietic Grafts' Potential," *Exp Hematol* 35 (2007): 1594–1600.
Baylis, Francoise, and Jason Scott Robert, "The Inevitability of Genetic Enhancement Technologies," *Bioethics* 18(1) (2004): 1–26.
Begley, Sharon, "U.S. Scientists Edit Genome of Human Embryo, but Cast Doubt on Possibility of 'Designer Babies'," *STAT*, August 2, 2017a: www.statnews.com/2017/08/02/crispr-designer-babies/
Begley, Sharon, "Tiny Human Brain Organoids Implanted into Rodents, Triggering Ethical Concerns," *STAT*, November 6, 2017b: www.statnews.com/2017/11/06/human-brain-organoids-ethics/
Begley, Sharon, "A Serious New Hurdle for CRISPR: Edited Cells Might Cause Cancer, Two Studies Find," *STAT*, June 11, 2018: www.statnews.com/2018/06/11/crispr-hurdle-edited-cells-might-cause-cancer/
Byrnes, W.M., "Why Human 'Altered Nuclear Transfer' Is Unethical: A Holistic Systems View," *Natl Cathol Bioeth Q* 5 (2005): 271–279.
Callaway, E., "Reproductive Medicine: The Power of Three," *Nature* 509 (7501) (2014): 414–417.

Center for Genetics and Society, "Report of First Gene-Edited Human Embryos in the US," July 27, 2017: www.geneticsandsociety.org/press-statement/report-first-gene-edited-human-embryos-us

Center for Genetics and Society, *In House Notes/Report*, September 2017.

Cha, Ariana Eunjung, "This Fertility Doctor Is Pushing the Boundaries of Human Reproduction, with Little Regulation," *Washington Post*, May 14, 2018: www.washingtonpost.com/national/health-science/this-fertility-doctor-is-pushing-the-boundaries-of-human-reproduction-with-little-regulation/2018/05/11/ea9105dc-1831-11e8-8b08-027a6ccb38eb_story.html

Chinnery, P.F., L. Craven, S. Mitalipov, J.B. Stewart, M. Herbert, and D.M. Turnbull, "The Challenges of Mitochondrial Replacement," *PLoS Genet* 10(4) (2014): e1004315.

CIRM, Memo/notes, "Before the Scientific and Medical Accountability Standards Working Group of the Independent Citizens' Oversight Committee to the California Institute for Regenerative Medicine Organized Pursuant to the California Stem Cell Research and Cures Act," February 4, 2016: www.cirm.ca.gov/sites/default/files/files/agenda/transcripts/Stds WkgGroup-2-4-16%20Transcript.pdf

Cohen, Jon, "'I Feel an Obligation to Be Balanced.' Noted Biologist Comes to Defense of Gene Editing Babies," *Science*, November 28, 2018: www.sciencemag.org/news/2018/11/i-feel-obligation-be-balanced-noted-biologist-comes-defense-gene-editing-babies

Connor, Steve, "Scientist Who Pioneered 'Three-Parent' IVF embryo Technique Now Wants to Offer It to Older Women Trying for a Baby," *The Independent*, February 7, 2015: www.independent.co.uk/news/science/three-parent-embryos-an-ivf-revolution-or-a-slippery-slope-to-designer-babies-10031477.html

Coughlin, S.M., "The Newman Application and the USPTO's Unnecessary Response," *Chi-Kent J Intell Prop* 5 (2006): 90–105.

Crabbe, J.C., D. Wahlsten, and B.C. Dudek, "Genetics of Mouse Behavior: Interactions with Laboratory Environment," *Science* 284 (1999): 1670–1672.

Cyranoski, D., "First Monkeys Cloned with Technique that Made Dolly the Sheep," *Nature* 553(7689) (2018): 387–388.

Darnovsky, Marcy, "Tired Tropes and New Twists in the Debate about Human Germline Modification," *Biopolitical Times*, May 28, 2015: www.geneticsandsociety.org/biopolitical-times/tired-tropes-and-new-twists-debate-about-human-germline-modification

Darnovsky, Marcy, Personal communication to TS, 2016.

Dawkins, Richard, "What's Wrong with Cloning?" Martha Craven Nussbaum and Cass R. Sunstein, eds, *Clones and Clones: Facts and Fantasies about Human Cloning*, New York: Norton, 1998, 54–66.

DeWitt, N., "Biologists Divided over Proposal to Create Human–Mouse Embryos," *Nature* 420 (2002): 255.

Doudna, Jennifer, "Genome-Editing Revolution: My Whirlwind Year with CRISPR," *Nature*, December 22, 2015: www.nature.com/news/genome-editing-revolution-my-whirlwind-year-with-crispr-1.19063?WT.ec_id=NEWSDAILY-20151222&spMailingID=50315178&spUserI D=MTM5MzQ0NTA2NjU1S0&spJobID=823388264&spReportId=ODIzMzg4MjY0S0

Dowie, Mark, "Gods and Monsters," *Mother Jones*, January–February, 2004: www.motherjones.com/politics/2004/01/gods-and-monsters/

Duncan, D.E., "Is the World Ready for Synthetic People?" *NEO.LIFE*, 2018: https://medium.com/neodotlife/q-a-with-drew-endy-bde0950fd038

Dunham-Snary, K.J., and S.W. Ballinger, "Anomalies Due to Incompatibilities Between Nuclei and Mitochondria that Have Not Coevolved," *Science* 349 (2015): 1449–1450.

Elliott, C. "Pharma Buys a Conscience," *The American Prospect* 12(17) (2001): 16–20.

Evans, N.G., and J.D. Moreno, "Children of Capital: Eugenics in the World of Private Biotechnology," *Ethics in Biology, Engineering and Medicine* 6 (2015): 285–297.

Farahany, Nita, "FDA Considers Controversial Fertility Procedure. What's at Stake?" *The Washington Post*, February 25, 2014: www.washingtonpost.com/news/volokh-conspiracy/wp/2014/02/25/fda-considers-controversial-fertility-procedure-whats-at-stake/?utm_term=.31560868ba0f

FDA *Guidance for Industry Document #187* (June 2015) Regulation of Genetically Engineered Animals Containing Heritable Recombinant DNA Constructs.

Fehilly, C.B., S.M. Willadsen, and E.M. Tucker, "Interspecific Chimaerism between Sheep and Goat," *Nature* 307(5952) (1984): 634–636.

Friend, Todd, "Silicon Valley's Quest to Live Forever," April 3, 2017: www.newyorker.com/magazine/2017/04/03/silicon-valleys-quest-to-live-forever

Fogarty, Norah M.E., Afshan McCarthy, Kirsten E. Snijders, Benjamin E. Powell, Nada Kubikova, Paul Blakeley, Rebecca Lea, Kay Elder, Sissy E. Wamaitha, Daesik Kim, Valdone Maciulyte, Jens Kleinjung, Jin-Soo Kim, Dagan Wells, Ludovic Vallier, Alessandro Bertero, James M. A. Turner, and Kathy K. Niakan, "Genome Editing Reveals a Role for OCT4 in Human Embryogenesis," *Nature* 550(7674) (2017): 67–73.

Gershoni, M., L. Levin, O. Ovadia, Y. Toiw, N. Shani, S. Dadon, N. Barzilai, A. Bergman, G. Atzmon, J. Wainstein, A. Tsur, L. Nijtmans, B. Glaser, and D. Mishmar, "Disrupting Mitochondrial-Nuclear Coevolution Affects OXPHOS Complex I Integrity and Impacts Human Health," *Genome Biol Evol* 6(10) (2014): 2665–2680.

Gould, Stephen Jay, "Individuality: Cloning and the Discomfiting Cases of Siamese Twins," *The Sciences* 37 (July–August 1997): 14–16.

Gould, Stephen Jay, "Dolly's Fashion and Louis's Passion," Martha Craven Nussbaum and Cass R. Sunstein, eds, *Clones and Clones: Facts and Fantasies about Human Cloning*, New York: Norton, 1998, 41–53.

Harkin, Tom, "Comments at Hearings of the Subcommittee on Public Health and Safety of the Senate Committee on Labor and Human Resources," March 12, 1997: www.cnn.com/HEALTH/9703/12/nfm/cloning/index.html

Hayes, Richard, "Human Genetic Engineering," Casey Walker, ed., *Made Not Born: The Troubling World of Biotechnology*, San Francisco, CA: Sierra Club Books, 2000, 80–98.

Heled, Y. "On Patenting Human Organisms or How the Abortion Wars Feed into the Ownership Fallacy," *Cardozo Law Review* 36 (2014): 241–298.

Hosseini, S.M., V. Asgari, S. Ostadhosseini, M. Hajian, H.R. Ghanaei, and M.H. Nasr-Esfahani, "Developmental Competence of Ovine Oocytes after Vitrification: Differential Effects of Vitrification Steps, Embryo Production Methods, and Parental Origin of Pronuclei," *Theriogenology* 83(3) (2015): 366–376.

Hubbard, R., and S.A. Newman, "Yuppie Eugenics," *Z Magazine*, March 1, 2002: https://zcomm.org/zmagazine/yuppie-eugenics-by-ruth-hubbard-and-stuart-newman/

Hurlbut, W.B., "Altered Nuclear Transfer: A Way Forward for Embryonic Stem Cell Research," *Stem Cell Rev* 1 (2005): 293–300.

Huxley, Aldous, *Brave New World, a Novel*, Garden City, NY: Doubleday Doran, 1932.

Hyun, Insoo, "Illusory Fears Must Not Stifle Chimera Research," *Nature* 537(7620) (September 13, 2016): www.nature.com/news/illusory-fears-must-not-stifle-chimaera-research-1.20582

Ishii, T., "Potential Impact of Human Mitochondrial Replacement on Global Policy Regarding Germline Gene Modification," *Reprod Biomed Online* 29(2) (2014): 150–155.

James, D., S.A. Noggle, T. Swigut, and A.H. Brivanlou, "Contribution of Human Embryonic Stem Cells to Mouse Blastocysts," *Dev Biol* 295(1) (2006): 90–102.

"Japan to Lift Ban on Growing Human Organs in Animals," *The Japan Times*, January 30, 2018: www.japantimes.co.jp/news/2018/01/30/national/science-health/japan-lift-ban-creating-human-organs-animals/#.W95RS_ZFyF1

Jasanoff, Sheila, and J. Benjamin Hurlbut, "A Global Observatory for Gene Editing," *Nature*, March 21, 2018: www.nature.com/articles/d41586-018-03270-w

Kaiser, Jocelyn, "NIH Moves to Lift Moratorium on Animal–Human Chimera Research," *Science*, August 4, 2016: www.sciencemag.org/news/2016/08/nih-moves-lift-moratorium-animal-human-chimera-research

Kelly, Mary Louise, "Harvard Medical School Dean Weighs In on Ethics of Gene Editing," *NPR, All Things Considered*, November 29, 2018: www.npr.org/2018/11/29/671996695/harvard-medical-school-dean-weighs-in-on-ethics-of-gene-editing

Khazan, Olga, "Babies Floating in Fluid-Filled Bags: A Lab Has Successfully Gestated Premature Lambs in Artificial Wombs. Are Humans Next?," *The Atlantic*, April 25, 2017: www.theatlantic.com/health/archive/2017/04/preemies-floating-in-fluid-filled-bags/524181/

King, David, "The Threat of Human Genetic Engineering," *Human Genetics Alert*, n.d.: www.hgalert.org/topics/hge/threat.htm

King, David, "Editing the Human Genome Brings Us One Step Closer to Consumer Eugenics," *The Guardian*, August 4, 2017: www.theguardian.com/commentisfree/2017/aug/04/editing-human-genome-consumer-eugenics-designer-babies

Kloc, M., R.M. Ghobrial, E. Borsuk, and J.Z. Kubiak, "Polarity and Asymmetry During Mouse Oogenesis and Oocyte Maturation," *Results Probl Cell Differ* 55 (2012): 23–44.

Lanphier, Edward, Fyodor Urnov, Sarah Ehlen Haecker, Michael Werner, and Joanna Smolenski, "Don't Edit the Human Germline," *Nature* 519 (2015): 410–411: www.nature.com/news/don-t-edit-the-human-germ-line-1.17111

Lavazza, A., and M. Massimini, "Cerebral Organoids: Ethical Issues and Consciousness Assessment," *J Med Ethics* 44(9) (2018): 606–610.

Lederberg, Joshua, "Experimental Genetics and Human Evolution," *The American Naturalist* 100(915) (Sep.–Oct., 1966): 519–531.

Ledford, Heidi, "Bitter CRISPR Patent War Intensifies," *Nature*, October 26, 2017: www.nature.com/news/bitter-crispr-patent-war-intensifies-1.22892

Liskovykh, M., S. Ponomartsev, E. Popova, M. Bader, N. Kouprina, V. Larionov, N. Alenina, and A. Tomilin, "Stable Maintenance of de Novo Assembled Human Artificial Chromosomes in Embryonic Stem Cells and Their Differentiated Progeny in Mice," *Cell Cycle* 14(8) (2015): 1268–1273.

Longaker, M.T., L.C. Baker, and H.T. Greely, "Proposition 71 and CIRM: Assessing the Return on Investment," *Nature Biotechnol* 25(5) (2007): 513–521.

Longmore Institute on Disability: https://longmoreinstitute.sfsu.edu

Lowthorp, Leah, "Opening the Door to Genetically Engineered Future Generations: How the NAS Report Ignores Widespread International Agreement," *Biopolitical Times*, Center for Genetics and Society, February 22, 2017: www.geneticsandsociety.org/biopolitical-times/opening-door-genetically-engineered-future-generations-how-nas-report-ignores

Magnani, T.A., "The Patentability of Human–Animal Chimeras," *Berkeley Tech Law J* 14 (1999): 443–460.

Meinecke-Tillmann, S., and B. Meinecke, "Experimental Chimaeras: Removal of Reproductive Barrier between Sheep and Goat," *Nature* 307 (5952) (1984): 637–638.

Merton, Robert K., *Social Theory and Social Structure*, New York: Free Press, 1968.

Myhrvold, N., "Human Clones: Why Not?" *Slate*, 1997: www.slate.com/articles/briefing/critical_mass/1997/03/human_clones_why_not.html

National Academies of Sciences, News: International Summit on Human Gene Editing, December 3, 2015: www8.nationalacademies.org/onpinews/newsitem.aspx?RecordID=12032015a

National Academies of Sciences, Engineering, Medicine, "Human Genome Editing: Science, Ethics, and Governance," The National Academies Press, 2017: www.nap.edu/read/24623/chapter/1

National Institutes of Health Guidelines on Human Stem Cell Research, 2009: http://stemcells.nih.gov/policy/pages/2009guidelines.aspx

Nature, editorial, "After Asilomar: Scientist-Led Conferences Are No Longer the Best Way to Resolve Debates on Controversial Research," 526 (October 15, 2015): 293–294: www.nature.com/news/after-asilomar-1.18546

Newman, Stuart A., "Cloning Our Way to 'the Next Level'," *Nature Biotechnology* 15 (1997): 488.

Newman, Stuart A., "Averting the Clone Age: Prospects and Perils of Human Developmental Manipulation," *Journal of Contemporary Health Law and Policy* 19(1) (2003): 431–463.

Newman, Stuart A., "My Attempt to Patent a Human–Animal Chimera," *L'observatoire de la génétique* 27, April–May, 2006: www.ircm.qc.ca/bioethique/obsgenetique/zoom/zoom_06/z_no27_06/za_no27_06_01.html

Newman, Stuart A., "Meiogenics: Synthetic Biology Meets Transhumanism," *GeneWatch* 25(1–2) (2012): 31, 37.

Newman, Stuart A., "Deceptive Labeling of Radical Embryo Construction Methods," *GeneWatch* 27(3) (November 21, 2014): http://issuu.com/genewatchmagazine/docs/genewatch_27-3

Newman, Stuart A., "CRISPR Will Never Be Good Enough to Improve People," *HuffPost*, February 22, 2017a: www.huffingtonpost.com/entry/crispr-will-never-be-good-enough-to-improve-people_us_58a90dcbe4b0b0e1e0e20c00

Newman, Stuart A., "Sex, Lies, and Genetic Engineering: Why We Must (But Won't) Ban Human Embryo Modification," I. Braverman, ed., *Gene Editing, Law, and the Environment: Life Beyond the Human*, London: Routledge, 2017b, 133–151.

NIH Guidelines on Human Stem Cell Research, 2009: http://stemcells.nih.gov/policy

NIH Research Involving Introduction of Human Pluripotent Cells into Non-Human Vertebrate Animal Pre-Gastrulation Embryos: http://grants.nih.gov/grants/guide/notice-files/NOT-OD-15-158.html

Obasogie, Osagie K., "Revisiting *Gattaca* in the Age of Trump," *Scientific American*, November 1, 2017: https://blogs.scientificamerican.com/observations/revisiting-gattaca-in-the-era-of-trump/

Parthasarathy, Shobita, *Patent Politics: Life Forms, Markets and the Public Interest in the United States and Europe*, Chicago, IL: University of Chicago Press, 2017.

Pei, D., D.W. Beier, E. Levy-Lahad, G. Marchant, J. Rossant, J.C. Izpisua Belmonte, R. Lovell-Badge, R. Jaenisch, A. Charo, and D. Baltimore, "Human Embryo Editing: Opportunities and Importance of Transnational Cooperation," *Cell Stem Cell* 21 (2017): 423–426.

Pence, Gregory, *Who's Afraid of Human Cloning?* London: Rowman & Littlefield, 1998.

Pereira, C.F., R. Terranova, N.K. Ryan, J. Santos, K.J. Morris, W. Cui, M. Merkenschlager, and A.G. Fisher, "Heterokaryon-Based Reprogramming of Human B Lymphocytes for Pluripotency Requires Oct4 but Not Sox2," *PLoS Genet* 4(9) (2008): e1000170.

Reardon, S., "Reports of Three-Parent Babies Multiply," *Nature*, October 19, 2016: www.nature.com/news/reports-of-three-parent-babies-multiply-1.20849

Regalado, A., "Human–Animal Chimeras Are Gestating on U.S. Research Farms," *MIT Technology Review*, January 6, 2016: www.technologyreview.com/s/545106/human-animal-chimeras-are-gestating-on-us-research-farms/

Regalado, A., "MIT Severs Ties to Company Promoting Fatal Brain Uploading," *MIT Technology Review*, April 3, 2018: www.technologyreview.com/s/610743/mit-severs-ties-to-company-promoting-fatal-brain-uploading/

Ridley, Matt, "Mitochondrial Donation Is a Wonderful Opportunity," *Rational Optimist*, 2015: www.rationaloptimist.com/blog/mitochondrial-donation-is-a-wonderful-opportunity/.

Robert, J., and F. Baylis, "Crossing Species Boundaries," *American Journal of Bioethics* 3(3) (2003): 1–13.

Roberts, Michelle, "Scientists Build 'Synthetic Embryos'," BBC News online, May 3, 2018: www.bbc.com/news/health-43960363

Sample, Ian, "Genetically Modified Babies Given Go Ahead by UK Ethics Body," *The Guardian*, July 27, 2018: www.theguardian.com/science/2018/jul/17/genetically-modified-babies-given-go-ahead-by-uk-ethics-body

Seok, J., H.S. Warren, A.G. Cuenca, M.N. Mindrinos, H.V. Baker, W. Xu, D.R. Richards, G.P. McDonald-Smith, H. Gao, L. Hennessy, C.C. Finnerty, C.M. Lopez, S. Honari, E.E. Moore, J.P. Minei, J. Cuschieri, P.E. Bankey, J.L. Johnson, J. Sperry, A.B. Nathens, T.R. Billiar, M.A. West, M.G. Jeschke, M.B. Klein, R.L. Gamelli, N.S. Gibran, B.H. Brownstein, C. Miller-Graziano, S.E. Calvano, P.H. Mason, J.P. Cobb, L.G. Rahme, S.F. Lowry, R.V. Maier, L.L. Moldawer, D.N. Herndon, R.W. Davis, W. Xiao, and R.G. Tompkins, "Genomic Responses in Mouse Models Poorly Mimic Human Inflammatory Diseases," *Proc Natl Acad Sci USA* 110(9) (2013): 3507–3512.

Shahbazi, M.N., A. Jedrusik, S. Vuoristo, G. Recher, A. Hupalowska, V. Bolton, N. N.M. Fogarty, A. Campbell, L. Devito, D. Ilic, Y. Khalaf, K.K. Niakan, S. Fishel, and M. Zernicka-Goetz, "Self-Organization of the Human Embryo in the Absence of Maternal Tissues," *Nat Cell Biol* 18(6) (2016): 700–708.

Sharma, Arun, Vittorio Sebastiano, Christopher T. Scott, David Magnus, Naoko Koyano-Nakagawa, Daniel J. Garry, Owen N. Witte, Hiromitsu Nakauchi, Joseph C. Wu, Irving L. Weissman, and Sean M. Wu, Letter to the Editor, "Lift NIH Restrictions on Chimera Research," *Science* 350(6261) (November 6, 2015): 640.

Shen, Helen, "The Labs Growing Human Embryos for Longer than Ever Before," *Nature* (July 4, 2018): www.nature.com/articles/d41586-018-05586-z

Silver, Lee, *Remaking Eden: Cloning and Beyond in a Brave New World*, New York: William Morrow, 1997.

Singer, Peter, "Shopping at the Genetic Supermarket," S.Y. Song, Y.M. Koo, and D.R.J. Macer (eds), *Asian Bioethics in the 21st Century*, Christchurch, New Zealand: Eubios Ethics Institute, 2003, 143–156.

Specter, Michael, "A Life of Its Own," *New Yorker*, September 28, 2009: 56–65.

Specter, Michael, "The Gene Hackers," *New Yorker Magazine*, November 16, 2015: www.newyorker.com/magazine/2015/11/16/the-gene-hackers

Stein, Rob, "Advance in Human Embryo Research Rekindles Ethical Debate," *Shots, Health News from NPR*, May 4, 2016: www.npr.org/sections/health-shots/2016/05/04/476539552/advance-in-human-embryo-research-rekindles-ethical-debate

Stein, Rob, "In Search for Cures, Scientists Create Embryos That Are Both Animal and Human," *NPR, All Things Considered*, May 18, 2016: www.npr.org/sections/health-shots/2016/05/18/478212837/in-search-for-cures-scientists-create-embryos-that-are-both-animal-and-human?sc=17&f=2&utm_source=iosnewsapp&utm_medium=Email&utm_campaign=app

Stein, Rob, "Exclusive: Inside the Lab Where Scientists Are Editing DNA in Human Embryos," *NPR, KQED Public Media*, August 18, 2017: www.npr.org/sections/health-shots/2017/08/18/543769759/a-first-look-inside-the-lab-where-scientists-are-editing-dna-in-human-embryos

Stein, Rob, "Inside the Ukrainian Clinic Making '3-Parent-Babies' for Women Who Are Infertile," *NPR, Minnesota Public Radio*, June 6, 2018: www.mprnews.org/story/2018/06/06/inside-the-ukrainan-clinic-making-three-parent-babies

Stevens, M.L. Tina, *The History of Bioethics: Its Rise and Significance*, San Francisco, CA: San Francisco State University, Elsevier Inc., 2014: https://pdfs.semanticscholar.org/411c/465786abd8240a3b6e2745df1c9338f22c79.pdf

Trounson, A., R. Klein, and R. Murphy, "Stem Cell Research in California: The Game Is On," *Cell* 132(4) (2008): 522–524.

UC Berkeley Events, "Jennifer Doudna and Sid Mukherjee in Conversation," January 26, 2018: www.youtube.com/watch?v=4fjwj92UNn4

Wadhwa, Vivek, "If You Could 'Design' Your Own Child, Would You?" *The Washington Post*, July 27, 2017: www.washingtonpost.com/news/innovations/wp/2017/07/27/human-editing-has-just-become-possible-are-we-ready-for-the-consequences/?utm_term=.50e17fbe6342

Weiss, Rick, "U.S. Ruling Aids Opponent of Patents for Life Forms," *Washington Post*, June 17, 1999.

Weiss, Rick, "U.S. Denies Patent for a Too-Human Hybrid," *Washington Post*, February 13, 2005.

Wolf, D.P., N. Mitalipov, and S. Mitalipov, "Mitochondrial Replacement Therapy in Reproductive Medicine," *Trends Mol Med* 21(2) (2015): 68–76.

Wolfe, Alexandra, "Jennifer Doudna: The Promise and Peril of Gene Editing," *The Wall Street Journal*, March 11, 2016: www.wsj.com/articles/jennifer-doudna-the-promise-and-peril-of-gene-editing-1457724836

Yamanaka, S., and K. Takahashi, "[Induction of Pluripotent Stem Cells from Mouse Fibroblast Cultures]," *Tanpakushitsu Kakusan Koso* 51(15) (2006): 2346–2351.

Zhang, John, Hui Liu, Shiyu Luo, Zhuo Lu, Alejandro Chávez-Badiola, Zitao Liu, Mingxue Yang, Zaher Merhi, Sherman J. Silber, Santiago Munné, Michalis Konstantinidis, Dagan Wells, Jian J. Tang, and Taosheng Huang, "Live Birth Derived from Oocyte Spindle Transfer to Prevent Mitochondrial Disease," *Reproductive Medicine Online* (2017): www.rbmojournal.com/article/S1472-6483(17)30041-X

Zimmer, Ben, "An Acronym Breaks Out of the Lab," *Wall Street Journal*, January 9–10, 2016: C4: www.wsj.com/articles/crispr-breaks-out-of-the-lab-1452181544

Zwerdling, D., "Humanimals. All Things Considered," National Public Radio, USA, April 5, 1998: www.npr.org/news/healthsci/indexarchives/1998/apr/980405.02.html

7
Concluding Reflections

In her influential work, *The Death of Nature*, historian of science Carolyn Merchant described the transformational paradigm shift of the seventeenth century's Scientific Revolution: "Rational control over nature, society, and the self was achieved by redefining reality itself through the new machine metaphor," Merchant explained (Merchant, 1980, p. 193). The animated, living nature of traditional thought and culture became dead, mechanized, and fully manipulable. Moreover, the process of this transformation involved wresting nature's secrets from "her." As Merchant extensively documents, "Female imagery became a tool in adapting scientific knowledge and methods to a new form of human power over nature" (Merchant, 1980, p. 165).

The paradigm shift of the late twentieth century's biotechnological revolution involves two great ironies pertaining to these two cultural aspects of the seventeenth century's scientific revolution. The first is that to control nature, "life" now needed to be resurrected. Crucial legal developments, *Diamond vs. Chakrabarty* and the Bayh-Dole legislation, speak to this revitalization. These twin 1980 measures rested on a shared conceptual foundation: profiting from nature now meant being able to own the ways its *living* processes could be directed to desired outcomes. Nature was no longer dead. It had just lost its agency. By the turn of the Millennium, it became clear that nature's new masters – a new breed of professionals, with feet planted equally in the laboratory and the boardroom – were targeting the human species for transformation. (In the process, practitioners and their allies sought to erase any real distinction between the natural and the artificial (Newman, 2010).) The second great irony is that the scientific revolution's consideration of nature as female and in need of having secrets wrested from her in order to command full control, came full circle: a key battleground of the biotechnological revolution is women's reproductive techno-strategies and a coveted raw material, women's eggs. Whereas the seventeenth century explicitly referenced female imagery, however, the role of women's bodies in the twentieth–twenty-first centuries' revolution is, as we have seen, too often concealed.

When, in 1997, developmental biologist Stuart Newman filed a patent for creating embryos that would be part human and part chimp, he wasn't concerned with protecting his intellectual property rights. Instead, he meant filing the patent to flag trends developing in biotechnological and reproductive science and practice. They were trends that compromised human biological integrity, and it was important for people to understand what was coming, he felt (Newman, 2006, and Chapter 6, this volume). Filing a preemptory patent was a compact strategy for delineating the issues, triggering public awareness, and focusing attention. Since then, controversial issues about the human genetic engineering associated with reproductive biotechnologies continue to be obscured by their promoters who promulgate misleading characterizations of their undertakings, sometimes engaging in outright deception.

Deception seems to be a recurring theme in the story of emerging biotechnologies. Two events in California, a global hub of biotechnological research and development, demonstrate this proclivity. The first is the 2004 passage of the California Stem Cell Research and Cures initiative. Managing to hide tendentious issues in plain sight, the new law resulted in a constitutional right to clone human embryos (Chapters 3 and 4). The second, less successful, was the campaign of the Lawrence Berkeley National Lab (LBNL) and UC Berkeley (UCB) to site a lab in the densely populated area of San Francisco's East Bay (Chapter 5). If successful it would have comprised a massive compound with contentious extreme genetic engineering applications a targeted focus. Strategies employed in both these efforts suggested the immense power, money, and influence of biotech boosterism, as well as a readiness to deceive. The earlier transformation of an urban belt in the San Francisco East Bay into a region ripe for developing controversial biotechnologies itself tells a story of stealth.

In 2007, a partnership between UC Berkeley's Chancellor, the Director of the Lawrence Berkeley National Laboratory, and a handful of local mayors dubbed a stretch of the San Francisco Bay Area's East Bay shoreline, "The East Bay Green Corridor" (hereafter, "Green Corridor"). The branding, part of the effort to "catch the green wave," as then Berkeley Mayor Tom Bates put it, voiced an aspiration to become the global hub of a putative emerging green economy. In 2009, local media reported on the federal government's grant of stimulus funds for the project, "for research, job training, weatherization and other environmentally themed projects" (Jones, 2009). In fact, of the over $76 million designated for the Green Corridor, only $17.4 million targeted Green Corridor cities. Most of the funds, a total of $58.5 million, were slated for two San Francisco East Bay institutional residents behind the Green Corridor creation: UC Berkeley and Lawrence Berkeley National Lab (known also as Berkeley Lab) (Brownstein, 2009). What was the public told NLRB would be doing? "[E]nergy research. Some of the research projects will use East Bay cities as a 'living lab' to study building energy efficiency" (Jones, 2009).

Two years later, declaring that Berkeley Lab had outgrown its space, LBNL and UCB announced the plans to create a "second campus" in one of the Green Corridor cities. No mention was made of LBNL's or UCB's roles in advancing controversial synthetic biology strategies or practices (Chapter 5). Nothing indicated the existence of swirling regulatory contentions concerning bio-safety, or how the intended lab park (described as a "second campus") would effectively create a synthetic biology epicenter amidst smoldering global dissension. The signed statement of 58 environmental, public interest, and religious groups from around the world was completely ignored. This concealed any indication of how the signatories considered synthetic biology, as practiced, to be decidedly not, "green." Nor was it clarified how key bioentrepreneur-faculty, with local bio-commercial ties, would benefit de facto by Green Corridor largesse and development. In 2011, the Sierra Club's San Francisco Bay Chapter offered comment on LBNL's process for choosing the eventual site. Its assessment included some hesitations about specific candidate sites. But its generally sanguine appraisal mirrored the Lab's avoidance in declaring synthetic biology a prime object of research and development – there was no reflection there (BondGraham, 2011). Ignoring controversy can be effective strategy.

Berkeley's Lawrence Hall of Science nestles on a ridge above the UCB campus, sited auspiciously near its cloistered neighbor, LBNL. Each day, its self-described mission "to inspire and foster learning for all" gets underway from an inviting hillside crux commanding a breathtaking viewscape of some of the cities, wetlands, and shoreline of the East Bay. Beyond this middle vision, the City of San Francisco rises into view, its Golden Gate passage to the world visible to the horizon. Educating begins wordlessly from this parking lot vista: there's always more than what you see. Few visitors recognize they are gazing over the East Bay Green Corridor, a region colonized by boosters for biotechnological research and development, much of it controversial. As biotechnology folds into general science education, what does science instruction teach about the field? What goes untaught?

In the summer of 2017, one of the Hall's galleries displayed arresting promotional panels signaling plans for a new exhibit, a "biotech learning lab." Larger than life photos of intent preschoolers manipulating beakers framed a quote from the president of biotech heavy-hitter Amgen, one of the Hall's major corporate donors. Positioned to connect the dots between a "Biotech Learning Lab" and "A World of Possibilities," the message explained, "While we must support those who will drive innovation in the future, we also need to deepen scientific understanding among all citizens – as they will be the ones to ultimately support it." Children will learn to drive and support the biotech world of possibilities. And like all the Hall's interactive exhibits, this learning lab should be an opportunity to make science fun and engaging.

Biotechnology, of course, is more than just a field of scientific study. Basic and applied science form a single technological platform where the desire for commercializable bio-products and patent opportunities often drive research agendas. So the Hall's planned exhibit raises questions not just about science but about social science, culture, and ethics, too. For example, what, besides science, will children learn about biotech? What is "the biotech world of possibilities"? "[C]loning, gene editing, DNA sequencing and more have revealed a world of possibilities," a second panel boasts, with scarcely a hint at the social and moral tumult riding in tandem with these biotechnologies. More than 40 countries around the globe have prohibitions against human germline modification. Even so, as we saw in Chapter 6, a report from the U.S. National Academies of Science and National Academy of Medicine now endorses what the New York Times characterized as "once unthinkable": clinical trials using heritable germline genome editing to create heritable traits.

Subsequently, Center for Genetics and Society Executive Director Marcy Darnovsky connected the dots biotech educators seem ready to ignore: The report acknowledges, she said, "many of the widely recognized risks, including stigmatizing people with disabilities, exacerbating existing inequalities, and introducing new eugenic abuses." But, she added, "[s]trangely, there's no apparent connection between those dire risks and the recommendation to move ahead." What better way to facilitate a future where citizens are rendered unable to detect connections between dire risks and bullish recommendations than to finance a Hall of Science that functions as a Corridor to Controversy Nullification?

When activists engaged events covered in this book, it was because they heard the silence of controversies nullified and with that the silenced possibilities for robust public awareness and civic input. They recognized that hiding the cloning of embryos meant hiding the need for women's eggs. It meant masking the increasing demand for young women to bear the understudied long-term health risks of egg extraction. They recognized that trumpeting the uses of synthetic biology concealed the need for vast amounts of biomass for feedstocks. It meant ignoring the devastating effect of that appetite on communities of the global south. Each time activists decided to take action, they were raising an implicit question: should bioentrepreneurial instincts be the ones forming our future?

A resounding, "NO" to this query is not an anti-science rejoinder. Biotechnology is not simply science. It is science in application. Bioentrepreneurs are not Galileos. Critics who call for civil oversight of biotechnological direction and application, especially as it pertains to modifying the human species, are not repressive authorities engaged in censorious ham-fisted meddling. Civil oversight is a matter of civic entitlement, perhaps even of civic duty. How can this entitlement be claimed, and the duty undertaken when bioentrepreneurs with financial stakes in the game hawk promises, overstate technological precisions, and ignore, understate, or hide the environmental, social, and human costs of doing so? From where can information be marshaled?

Most journalists and bioethicists have so far appeared largely unable to assume a critical stance toward biotech. Instead they take at face value simplistic business models that don't stand up to relevant scientific scrutiny. Under what often seems to be a reflexive granting of the benefit-of-the-doubt, news coverage and commentary typically has offered up a pre-bundled formula for assessing the technology *de jour*: quote a critic, often mouthing a trope about the untowardness of "playing God," in an otherwise laudatory account; do not reflect on how the demise of genetic determinism means no one can be any good at playing God and how, already, some bear the insupportable risks of trying to do so. The package has proven to be a powerful normalizing force.

Another force, neutralizing dissent, involves actually inviting critics to the table. Once seated, they can acclimate to the heat of biotech trends intensifying despite unaddressed controversies. The authors of *Evolving Ourselves: Redesigning the Future of Humanity – One Gene at a Time* are co-founders of Excel Venture Management, "which builds start-ups in synthetic biology, big data, and new genetic technologies." After announcing the inevitability of redesigned humans, they invite the public to help get us there. "Humans are on the way to becoming something else. Something of their own design," they declare. "We are trying out various temporary solutions and recommendations before making permanent alternations. But the direction is clear." Their suggestion for how to have a "sensible conversation" about "rapid unnatural evolution" is to ask for, "an ongoing ethics and permitted uses conversation." "[H]elp us develop a set of guiding principles," they invite (Enriquez and Gullans, 2016, pp. 262–264). Then, they frame the intended dialogue by feeding readers suggestions about what to ask for on the road to their biotech destinations. There is no "whether" we will get there. Best Practices for arriving at the inevitable is all that can be hoped for. There is no stopping a juggernaut.

Or is there?

Is it possible to avoid succumbing to bio-mesmerism or co-optation? We hope so and that is why we have written this book. Stellar information-gathering civil society groups largely succeed in resisting them (Appendix I). A few manage to engage activism through petition drives, litigation, legislation, public forums, etc. But, with resources scarce, more husband their time, talent, and treasure by mastering facts and making connections in the hopes of inducing awareness and igniting action in others. Consulting them may provide the muscle needed to help move controversial issues out from behind the shuttered doors of meetings sequestered in the halls of academe or corporate backrooms, and off the tables of bioethics commissions. The public is inoculated against insight into plans hatched or parsed in these venues. In the town square, armed with facts and with biotechnology in hand, citizens may perhaps secure the critical mass necessary to lead us to a humane and human future.

Biotech is a juggernaut, only if allowed to be.

Sources Consulted for Chapter 7

BondGraham, Darwin. "The Race for Berkeley Lab Nears Finish," *East Bay Express*, October 19, 2011: www.eastbayexpress.com/oakland/the-race-for-berkeley-lab-nears-finish/Content?oid=3019215

Brownstein, Zelda, "East Bay Green Corridor: Green for All?", *BeyondChron*, July 7, 2009: www.beyondchron.org/east-bay-green-corridor-green-for-all/

Enriquez, Juan, and Steve Gullans, *Evolving Ourselves: Redesigning the Future of Humanity – One Gene at a Time*, New York: Current, 2016.

Jones, Carolyn, "East Bay Green Corridor Grows, Cash Pours in," *SF Gate*, June 27, 2009: www.sfgate.com/business/article/East-Bay-Green-Corridor-grows-cash-pours-in-3294442.php

Merchant, Carolyn, *The Death of Nature: Women, Ecology, and the Scientific Revolution*, San Francisco, CA: Harper Row, 1980.

Newman, Stuart A., "My Attempt to Patent a Human–Animal Chimera," *L'observatoire de la génétique* 27 (April–May, 2006): www.ircm.qc.ca/bioethique/obsgenetique/zoom/zoom_06/z_no27_06/za_no27_06_01.html

Newman, Stuart A., "Renatured Biology: Getting Past Postmodernism in the Life Sciences," Cabell King and David Albertson, eds, *Without Nature?: A New Condition for Theology*, New York: Fordham University Press, 2010.

Appendices

Appendix A: Position Paper on Human Germ Line Manipulation Presented by Council for Responsible Genetics, Human Genetics Committee, Fall 1992

Appendix B: Memo to CIRM Standards Working Group Challenging CIRM's 2013 Move to Change Egg Donation Reimbursement Policy

Appendix C: Testimony of Sindy Wei, MD. to the California Senate Health Committee re AB 926, June 12, 2013[1]

Appendix D: Testimony of Jennifer Schneider, MD. to the California Senate Health Committee re AB 926 June 12, 2013[2]

Appendix E: Testimony of Raquel Cool, Co-Founder of We Are Egg Donors, to the California Assembly Health Committee re AB 2531, April 5, 2016

Appendix F: Email to SFSU Faculty from UCSF Center for Reproductive Health, September 21, 2017

Appendix G: Open Letter to President's Bioethics Commission from Fifty-Eight Civil Society Groups, December 16, 2010[3]

Appendix H: Civil Society Letter to East Bay City Councils, September 27, 2011

Appendix I: List of Civil Society Organizations Tracking Emerging Biotechnologies

Notes

1 See: www.geneticsandsociety.org/internal-content/testimony-sindy-wei-md-california-senate-health-committee-re-ab-926
2 See: www.geneticsandsociety.org/internal-content/testimony-jennifer-schneider-md-california-senate-health-committee-re-ab-926
3 See: www.etcgroup.org/sites/www.etcgroup.org/files/publication/pdf_file/Civil%20Society%20Letter%20to%20Presidents%20Commission%20on%20Synthetic%20Biology_0.pdf

Appendix A
Position Paper on Human Germ Line Manipulation

Presented by Council for Responsible Genetics,
Human Genetics Committee Fall, 1992

THE POSITION OF THE COUNCIL FOR RESPONSIBLE GENETICS

The Council for Responsible Genetics (CRG) strongly opposes the use of germ line gene modification in humans. This position is based on scientific, ethical, and social concerns.

Proponents of germ line manipulation assume that once a gene implicated in a particular condition is identified, it might be appropriate and relatively easy to change, supplement or otherwise modify the gene by some form of therapy. However, biological characteristics or traits usually depend on interactions among many genes, and these genes are themselves affected by processes that occur both inside the organism and in its surroundings. This means that scientists cannot predict the full effect that any gene modification will have on the traits of people or other organisms. In purely biological terms, the relationship between genes and traits is not well enough understood to guarantee that by eliminating or changing genes associated with traits one might want to avoid, we may not simultaneously alter or eliminate traits we would like to preserve. Even genes that are associated with diseases that may cause problems in one context can be beneficial in another context.

Two frequently destructive aspects of contemporary culture are linked together in an unprecedented fashion in germ line gene modification. The first is the notion that the value of a human being is dependent on the degree to which he or she approximates some ideal of biological perfection. The second is the ideology that all limitations imposed by nature can and should be overcome by technology. To make intentional changes in the genes that people will pass on to their descendants would require that we, as a society, agree on how to identify 'good' and 'bad' genes. We do not have such criteria, nor are there mechanisms for establishing them. Any formulation of such criteria would necessarily reflect current social baises.

Moreover, the definition of the standards and the technological means for implementing them would largely be determined by the economically and socially privileged. By implementing a program of germ line manipulation these groups would exercise unwarranted influence over the common biological heritage of humanity.

What Is "Germ Line Manipulation"?

The undifferentiated cells of an early embryo develop into either germ cells or somatic cells. *Germ* cells, or reproductive cells, are those that develop into the egg or sperm of a developing organism and transmit all its heritable characteristics. *Somatic* cells, or body cells, refer to all other cells of the body. While both types of cells contain chromosomes, only the chromosomes of germ cells are passed on to future generations.

Techniques are now available to change chromosomes of animal cells by inserting new segments of DNA into them. If this insertion is performed on specialized or *differentiated* body tissues, such as liver, muscle, or blood cells, it is referred to as *somatic cell* gene modification, and the changes do not go beyond the individual organism. If it is performed on sperm or eggs before fertilization, or on the undifferentiated cells of an early embryo, it is called *germ cell* or *germ line* gene modification, and the changes are not limited to the individual organism. For when DNA is incorporated into an embryo's germ cells, or undifferentiated cells that give rise to germ cells, the introduced gene or genes will be passed on to future generations and may become a permanent part of the gene pool.

Deliberate gene alterations in humans are often referred to as 'gene therapy.' The Council for Responsible Genetics (CRG) prefers to use the terms 'gene modification' and 'gene manipulation' because the word 'therapy' promises health benefits, and it is not yet clear that gene manipulations are beneficial.

Why Might Germline Modification Be Attempted in Humans?

If one or both partners carry a version of a gene that could predispose their offspring to inherit a condition they want to avoid, genetic manipulation may appear to be a potential way to prevent the undesired outcome. The earlier during embryonic development the targeted gene or genes are replaced, the less likely is the resulting individual to be affected by the unwanted gene. But while the immediate goal of such a modification might be to alter the genetic constitution of a single individual, modifications made at the early embryonic stages would incidentally result in germ line modification, and so all the offspring of this person would have and pass on the modification.

Alternatively, germ line modification may be the intended consequence of the procedure. One goal might be to 'cleanse' the gene pool of 'deleterious' genes. For example, Daniel E. Koshland, Jr., a molecular biologist, and the editor-in-chief of *Science*, has written, "keeping diabetics alive with insulin, which increases the propagation of an inherited disease, seems justified only if one ultimately is willing to do genetic engineering to remove diabetes from the germ line and thus save the anguish and cost to millions of diabetics." (1) Another goal of germ line manipulation may be to avoid multiple treatments of somatic gene modification that would be required under proposed treatment protocols for certain conditions such as cystic fibrosis.

Some people may also look forward to the possibility of introducing genes into the germ line that can 'enhance' certain characteristics desired by parents or other custodians of the resulting offspring. In the article referred to above, Koshland raises the possibility that germ line alterations could be perceived to meet future 'needs' to design individuals "better at computers, better as musicians, better physically."

The attempt to improve the human species biologically is known as *eugenics*, and was the basis of a popular movement in Europe and North America during the first half of this century. Eugenics was advocated by prominent scientists across the entire political spectrum, who represented it as the logical consequence of the most advanced biological thinking of the period. In the U.S., eugenic thinking resulted in social policies that called for forced sterilization of individuals regarded as inferior because they were 'feeble minded or paupers.' In Europe, the Nazis took up these ideas, and their attempts at implementation led to widespread revulsion against the concept of eugenics. Today public discussion in favor of influencing the genetic constitution of future generations has gained new respectability with the increased possibility for intervention presented by in-vitro fertilization and embryo implantation technologies. Although it is once again espoused by individuals with a variety of political perspectives, the doctrine of social advancement through biological perfectibility underlying the new eugenics is almost indistinguishable from the older version so avidly embraced by the Nazis.

It is important to recognize that the dream of eliminating 'harmful' genes (such as those associated with cystic fibrosis or Duchenne muscular dystrophy) from the entire human gene pool could be realized only over time scales of thousands of years, and then only with massive, coercive programs of germ line manipulation. Such a program would be neither feasible nor morally acceptable. As a practical matter then, any presumed beneficial effects of germ line modification would pertain to individual families, not to the human population as a whole. This is in contrast to harmful effects, which would be widely disseminated.

Furthermore, parents who carry a gene which they would not want a child of theirs to inherit could arrange to have unaffected, biologically related offspring *without* germ line modification. If a gene is well enough characterized to consider gene manipulation, there will always be a diagnostic test available to identify a fetus that carries that gene and parents, if they choose, may then terminate the pregnancy. Given that there are alternatives for avoiding the inheritance of unwanted genes, the main selling point of germ line modification techniques over the long term would appear to be the prospect of enhancement of desired traits.

What Is the Feasibility of Modifying the Germline of Humans?

Both somatic and germ line modification are widely performed on laboratory animals for research purposes. Somatic gene modifications have already been performed on humans and additional experimental protocols are being approved by the National Institutes of Health in increasing numbers.

No published reports have yet appeared on germ line modification in humans, but there appear to be no technical obstacles to such experiments, and articles proposing these procedures are becoming more and more common in the literature (2, 3, 4). Germ line gene modification has actually proved technically easier than somatic modification in mice and other vertebrate animals which have been employed as 'models' for human biology in the past, because the cells of early embryos incorporate foreign DNA and synthesize corresponding functional proteins more readily than most differentiated somatic cells. A widely reported example of the successful experimental use of the germ line technique was the introduction of an extra gene that specified growth hormone into fertilized mouse eggs. In the presence of the high levels of growth hormone produced, the mice grew to double their normal size. Germ line techniques are also being used in attempts to modify farm animals, with stated goals of increasing yields or enhancing nutritional quality of meat and other animal products.

Given what has been accomplished in animals, the only remaining technical requirements for germ line gene modification in humans are procedures for collecting a woman's eggs, fertilizing them outside her body, and implanting them in the uterus of the same or another woman, where they can be brought to term. These are already well established procedures for humans and are widely used in *in-vitro* fertilization clinics.

What Are the Technical Pitfalls?

Current methods for germ line gene modification of mammals are inefficient, requiring the microinjection of numerous eggs with foreign DNA before an egg is successfully modified. Moreover, introduction of a foreign gene (even if there is a copy of one already present) into an inappropriate location in an embryo's chromosomes can have unexpected consequences. For example, the offspring of a mouse that received an extra copy of the normally present *myc* gene developed cancer at 40 times the rate of the unmodified strain of mice. (5)

Techniques to introduce foreign DNA into eggs, however, are constantly being improved and eventually will be portrayed as efficient and reliable enough for human applications. It may soon be possible to place a gene into a specified location on a chromosome while simultaneously removing the unwanted gene. This will increase the accuracy of the procedures, but does not eliminate the possibility that gene combinations will be created that will be harmful to the modified embryo, and its descendants in future generations. Such inadvertent damage could be caused by technical error, or more importantly, by biologists' inability to predict how genes or their products interact with one another and with the organism's environment to give rise to biological traits. It would have been impossible to predict, a priori, for example, that someone who has even *one* copy of the gene for a blood protein known as hemoglobin-S would be protected against malaria, whereas a person who has *two* copies of this gene would have sickle cell disease.

This unpredictability applies with equal force to genetic modifications introduced to 'correct' presumed disorders and to those introduced to enhance characteristics. Inserting new segments of DNA into the germ line could have major, unpredictable consequences for both the individual and the future of the species that include the introduction of susceptibilities to cancer and other diseases into the human gene pool.

What Are the Social and Ethical Implications of Germ Line Modification?

Clinical trials in humans to treat Adenosine Deaminase Deficiency—a life threatening immune disorder—and terminal cancer with somatic gene modification are already in progress and experiments to treat diabetes and hypertension are under development. It is important to distinguish the ethical problems raised by these protocols from the additional, and more profound questions raised by germ line modification. While the biological effects of somatic manipulations reside entirely in the individual in which they are attempted, such treatments are not strictly analogous to other therapies with individual risk. Radiation, chemical or drug treatment can be withdrawn if they prove harmful to the patient, while some forms of somatic modification cannot. Thus, somatic gene modification requires a person to forfeit his/her rights to withdraw from a research study because the intervention cannot be stopped, whether harmful or not. Valid objections have also been raised to the fact that the first somatic gene modification experiments, involving Adenosine Deaminase Deficiency, were carried out on young children who were not themselves in a position to give informed consent. While it appears that somatic gene modification techniques will be used increasingly in the future, the CRG urges that they be used with greatest caution, and only for clearly life-threatening conditions.

Germ line modification, in contrast, has not yet been attempted in humans. The Council for Responsible Genetics opposes it unconditionally. Ethical arguments against germ line modification include many of those that pertain to somatic cell modification, as well as the following:

- Germ line modification is not needed in order to save the lives or alleviate suffering of existing people. Its target population are 'future people' who have not yet even been conceived.
- The cultural impact of treating humans as biologically perfectible artifacts would be entirely negative. People who fall short of some technically achievable ideal would increasingly be seen as 'damaged goods.' And it is clear that the standards for what is genetically desirable will be those of the society's economically and politically dominant groups. This will only reinforce prejudices and discrimination in a society where they already exist.
- Accountability to individuals of future generations who are harmed or stigmatized by wrongful or unsuccessful germ line modifications of their ancestors is unlikely.

In conclusion, the Council calls for a ban on germ line modification.

References

1. Koshland Jr., Daniel E., "The future of Biological Research: What Is Possible and What Is Ethical?", *MBL Science*, v. 3, no. 2, pp. 11–15, 1988.
2. Walters, Leroy, "Human Gene Therapy: Ethics and Public Policy," *Human Gene Therapy*, v. 2, pp. 115–122, 1991.
3. Working Group on Genetic Screening and Testing, *Report of Discussions in Genetics, Ethics and Human Values*, XXIVth CIOMS Conference, Tokyo and Inuyama, Japan, 24–26 July 1990.
4. Buster, John E., and Carson, Sandra A., "Genetic Diagnosis of the Preimplantation Embryo," *American Journal of Medical Centers*, v. 34, pp. 211–216, 1989.
5. Leder, A. et al, "Consequences of Widespread Deregulation of the c-myc Gene in Trangenic Mice: Multiple Neoplasms and Normal Development," *Cell*, v. 42, p. 485, 1986.

This document was written by the Human Genetics Committee of the Council for Responsible Genetics (CRG). The Council is a Cambridge-based national organization of scientists, public health advocates, trade unionists, women's health activists and others who want to see biotechnology developed safely and in the public interest. The Council believes that an informed public can and should play a leadership role in setting the direction for emerging technologies. A fundamental goal of the CRG is to prevent genetic discrimination.

The Human Genetics Committee has 14 members with backgrounds in the biological sciences, public health, law, disability rights, occupational health and safety, and women's health. Members include: Abby Lippman, Professor of Epidemiology, McGill University, Chairperson; Philip Bereano, Professor of Engineering and Public Policy, University of Washington; Paul Billings, Chief of Genetic Medicine, Pacific Presbyterian Medical Center; Colin Gracey, Head of the Religious Life Office, Northeastern University; Mary Sue Henifin, Deputy Attorney General, State of New Jersey; Ruth Hubbard, Professor Emerita of Biology at Harvard University; Sheldon Krimsky, Associate Professor of Urban and Environmental Policy, Tufts University; Richard Lewontin, Alexander Agassiz Professor of Zoology, Harvard University; Karen Messing, Professor of Biology, University of Quebec in Montreal; Stuart Newman, Professor of Cell Biology and Anatomy, New York Medical College; Judy Norsigian, Co-Director, Boston Women's Healthbook Collective; Marsha Saxton, Director, Project on Women and Disability; Doreen Stabinsky, California Biotechnology Action Council and University of California at Davis; and Nachama L. Wilker, Executive Director, Council for Responsible Genetics.

The Council for Responsible Genetics, 19 Garden Street, Cambridge, MA 02138, USA. Telephone: 617-868-0870. Telefax: 617-864-5164.

Appendix B

Memo to CIRM Standards Working Group Challenging CIRM's 2013 Move to Change Egg Donation Reimbursement Policy

To: CIRM Standards Working Group
From: Center for Genetics and Society
Pro-Choice Alliance for Responsible Research
Alliance for Humane Biotechnology
Our Bodies Ourselves
Date: July 24, 2013
Re: SWG should reject the proposal to allow use of cell lines created with paid-for eggs

Summary of Problems with the Proposed Regulation Regarding Use of Stem Cell Lines

1. Procedural problems with the proposed policy. 1) The purpose of regulations is to give clear guidance to the public and affected parties – the proposed regulation does not meet that standard; 2) the stated criterion of "advancing CIRM's mission" for approving otherwise non-acceptable cell lines is arbitrary and much too vague; 3) The new procedure could require the SWG to abdicate its role in establishing medical and ethical standards as the exception could easily swallow the rule, and the SWG would no longer have a role in evaluating proposed exemptions.

2. Procedural problems with the decision-making process for this proposal. The SWG is being asked to make a weighty change with minimal discussion. The information and documentation provided to the SWG – and to interested members of the public – are woefully insufficient to allow an informed decision.

3. The justifications for the policy change are unconvincing. Many of the concerns that motivated the 2006 decision to prelude CIRM grantees from using cell lines created with paid-for eggs remain unresolved. For example, there is little publicly available information about the New York program that pays women for providing eggs for research.

4. Numerous concerns raised and risks acknowledged in "Proposed oocyte donation guidelines for stem cell research" (*Fertility and Sterility*, Dec 2010), prepared at the request of CIRM and co-authored by CIRM staff, do not appear to have been met.

5. The example provided of a cell line that would be allowed under this policy change is problematic: The OHSU team is presented as using ethically robust protocols and procedures. But the group's published papers acknowledge that they retrieved large numbers of eggs from some of the women they recruited – numbers larger than what CIRM's guidelines recommend.

Additional Information

1. Procedural problems with the proposed policy

- According to the memo from CIRM staff and the proposed regulatory amendment, the ICOC will decide whether to approve a grantee's request to use a stem cell line created with paid-for eggs on the basis of whether doing so "will advance CIRM's mission." This criterion is much too vague, and doesn't include consideration of the health or welfare of the women who undergo egg retrieval. Protecting the well-being of women providing eggs is not even mentioned (though perhaps it could be considered as an element of the fifth of five "factors to be considered by the ICOC," "whether the donation . . . was consistent with 'best practices' at the time of donation").

- The proposed procedure seems to take decisions about ethical standards including those pertaining to egg provider recruitment and egg retrieval protocols out of the purview of the SWG. The proposal states that "CIRM" will conduct "review and analysis" and "scientific and ethical evaluation." It does not state who at CIRM would conduct these studies, or what if any opportunity for review and input SWG would have. If the "evaluation" described in Appendix 1 of the July 1 briefing memo is indicative, the SWG (or whomever at CIRM makes the recommendation to the ICOC) might be told very little about such considerations. See below.

- One of the SWG's functions as provided in Proposition 71 is to "recommend to the ICOC standards for safe and ethical procedures for obtaining materials and cells for research and clinical efforts." If the proposed policy change is enacted, the SWG would not be in a position to fulfill this responsibility.

2. Procedural problems with the decision-making process for this proposal

- Proposals to allow women to be paid for eggs to be used as research materials have been in front of the Standards Working Group multiple times over the past 8 + years. It is a contentious and complex issue. Now the SWG is being asked to overturn policy that was set in 2006, with the support of then-president Zach Hall. This is an important decision that will have national and international significance. But the background information and documentation provided to the SWG (and to interested members of the public) are woefully insufficient as a basis for making this decision.
- Regarding the OHSU SCNT derivation, which is apparently a key factor motivating the proposal, the memo from CIRM staff says there was a CIRM site visit at OHSU that included reviews, interviews, discussions and evaluations. But each is described in a single phrase. There is no substantive information about any of the elements. For example, there is no information about the expertise of the ESCRO, the donor recruitment procedures, or the "specific steps taken to support participant safety and autonomy." Informed consent documents are said to have been reviewed, but are neither described nor provided.

3. The justifications for the policy change are unconvincing

- According to the memo from CIRM staff, at the time of the 2006 decision to preclude CIRM grantees from using stem cell lines where financial compensation was provided to egg providers, "there was scientific uncertainty concerning the feasibility and value off SCNT experiments and limited experience with oocyte donation programs for research." The memo goes on to list five "scientific and policy developments" since 2006 that presumably speak to the scientific uncertainty and limited experience with programs that pay women for eggs. But there is no discussion of whether, or in what regard, these developments in fact justify the policy change.
- Three of the five listed developments are scientific in nature. Of these, one – "the successful derivation of stem cell lines in New York and Oregon utilizing SCNT methodologies" – is directly relevant to the proposed policy change. The other two, involving iPSC and HESC-derived investigations and therapies, appear to be relevant only because the SCNT-derived lines would provide an additional cell type for comparison purposes.

- Not all scientists agree that it is necessary to compare stem cell lines derived via SCNT with iPS or HESC lines. For example, Alexander Meissner, a developmental biologist at the Harvard Stem Cell Institute in Cambridge, Massachusetts, says "Mitalipov's cell lines will not reveal much about how stem cells transform." "US scientists chafe at restrictions on new stem-cell lines: California centre rethinks rules in wake of discovery," *Nature*, June 4, 2013.

- Two of the five listed developments are policy matters. One of these is "development of CIRM oocyte donation guidelines." However, the concerns raised by these guidelines have not been adequately addressed by current practices in Oregon, in New York, or in the context of third-party paid egg retrieval for fertility purposes. See below.

- The second of the two policy developments listed is the establishment of "a paid oocyte donation program in New York." However, no information or documentation is provided about this program – and in fact our research has not found any that has been published. So it is impossible to know how to evaluate the New York program, and it certainly can't be taken as an argument in favor of the proposed policy change.

- One of the frequently cited justifications for paying women to provide eggs for research is that without payments beyond reimbursement for expenses, researchers cannot obtain the eggs they need. However, in reviewing the available information about New York's discussion of paying women to provide eggs for research, we noted in an article by Robert Klitzman and Mark Sauer ("Payment of egg donors in stem cell research in the USA," *Reproductive BioMedicine Online*, March 20, 2009) the following result from their survey of 230 women "enrolled or presently participating as egg donors in the Columbia University programme for assisted reproduction, who now receive an US$8000 payment for this service."

  o One of the questions posed was "Financial compensation for egg donors may be limited when they are donating for research purposes. If payment was limited to travel reimbursement only, would you still consider donating your eggs?" 43% of the respondents said yes; 6% were unsure.

  o Current California and CIRM policy permits women to be reimbursed for travel, housing, child care, medical care, health insurance, and lost wages.

4. Numerous concerns raised and risks acknowledged in "Proposed oocyte donation guidelines for stem cell research" (2009), co-authored by CIRM staff, do not appear to have been met

- Concerns that do not appear to have been met, and that therefore argue for maintaining the current policy that disallows use of stem cell lines derived using paid oocytes.
  - "Oocyte retrieval involves potential acute and long-term risk to the donor."
  - "Although oocyte donation has been performed for decades in the context of fertility treatments, donation for research will neither achieve pregnancy nor create direct therapeutic benefits to the donor or potential patients at this time."
  - "The IOM workshop and report discussed the unique ethical context in which oocyte donation for research exists – women incur some medical risk without direct benefit to themselves or others."
  - "The committee concurs with the observation in the IOM report that the absence of registries to track the health of oocyte donors represents a limitation for evaluating any long-term effects. There is a need for additional data that would be applicable to the population in question – ostensibly healthy donors who are not intending to undergo IVF at the time of donation."

- Increasing evidence of the significant risks of egg retrieval: While there has been no systematic follow-up of women who have undergone egg retrieval, we continue to hear from women who have been seriously harmed. See, for example:
  - "Paying women to take big risk" by Leah Campbell, *San Diego Union Tribune,* Jul 11, 2013
  - Testimony by Sindy Wei, MD to the California Senate Health Committee re AB 926, June 12, 2013
  - Testimony by Jennifer Schneider, MD to the California Senate Health Committee re AB 926, June 12, 2013
  - "I Donated My Eggs So Someone Else Could Get Pregnant," by Raquel Cool, May 30, 2013.

5. A problematic aspect of the SCNT derivation at OHSU

- "Proposed oocyte donation guidelines for stem cell research" recommends that one of the indicators for stopping an egg retrieval cycle is hyper-response, since this significantly increases the chances of more serious ovarian hyperstimulation syndrome. Hyper-response is defined as "> 20 follicles on day 6 of stimulation." In other words, retrieving > 20 eggs in a cycle should not happen: if that many egg follicles are ripening, the cycle should be cancelled.

- The CIRM memo asking for approval of the policy change says that CIRM's site visit at OHSU found that all aspects of OHSU's protocols were consistent with CIRM's policies.

- However, two recent scientific papers published by the OHSU research team, including the one on the SCNT derivation of stem cell lines, indicate that more than 20 eggs were retrieved from some egg providers. The papers do not give full information about the numbers of eggs retrieved per cycle, but it acknowledges that 28 eggs were retrieved from a woman in one study, and that in the SCNT study a *mean* of 20.5 eggs were retrieved from a small group of women (the number of women in the group isn't given, but it appears to be either 4 or 6), with a (surprisingly large) standard deviation of 11.9. From the articles:

- "Human Embryonic Stem Cells Derived by Somatic Cell Nuclear Transfer," Tachibana et al., *Cell 153, 1–11*, June 6, 2013: In its discussion of its "effort to define optimal stimulation protocols that are compatible with high-quality oocytes," it says: "The average number of MII oocytes (mean +- SD) collected per cycle" in two groups of egg providers was "11.7 +- 5.6" and "20.5 +- 11.9".

- "Towards germline gene therapy of inherited mitochondrial diseases," Tachibana et al., *Nature*, date: "a total of 106 . . . oocytes were retrieved (range of 7–28)."

Appendix C

Testimony by Sindy Wei, MD to the California Senate Health Committee re AB 926

Sindy Wei, By MD

My name is Sindy Wei. I am an M.D. with a Ph.D. in Biology, and I am a former egg donor. I have published in scientific journals including *Nature* and *Science*. I am a supporter of stem cell research.

When I heard this bill had been introduced into the CA legislature I assumed that egg donors finally would be given the protections of research subjects – the most basic of which is attention to the effects of the intervention on their health. I was disappointed to find that AB 926 only serves the interests of those who want the eggs, not the egg providers.

In 2001, I signed up for egg donation after researching medical literature. I injected hormones for many days. Early on I expressed concerns about the large numbers of egg follicles seen on my ultrasound, but doctors reassured me that this was great news. Then, before the retrieval, my blood estrogen levels rose much higher than anticipated. They decided to continue. The next morning, I underwent transvaginal needle retrieval of approximately 60 eggs.

I woke from the anesthesia feeling weak, nauseous, and short of breath. They told me I was ready to go home but I could not stand. After 8 hours of encouraging me to go home they FINALLY admitted me to the hospital.

Soon it became undeniable that I was going into shock from blood loss. I was taken to the operating room for emergency surgery and blood transfusion. Had I trusted their judgment one last time and gone home, I would have died.

After surgery, I had to be kept in the ICU. When the egg retrieval doctor came to see me she suggested that the bleeding was due to a genetic bleeding disorder (that is, my own fault). Testing revealed no such thing.

I was shocked by this dismissive attitude from a doctor of a top fertility treatment center, who has published articles on safety evaluation and recommendations for egg harvesting.

I fear that cases like mine are buried deep by fertility centers concerned about their image. An industry thriving on profits and reputation has little incentive to report adverse events, or protect the health and medical rights of donors.

Later, I developed unexplained infertility and had to be treated with even more hormones and surgeries. I still worry about the long-term risk of cancer. Please don't expand the market in human eggs unless minimal protections for egg donors are ensured, especially the long term follow up necessary to make genuine informed consent possible.

Appendix D

Testimony by Jennifer Schneider, MD to the California Senate Health Committee re AB 926

Jennifer Schneider, By MD

My name is Jennifer Schneider. I'm an Internal Medicine physician and the mother of a Stanford student egg donor. I would support AB 926 if it did what its authors claimed, but it does not.

I'm concerned about increasing the number of women who go through egg donation because of what happened to my daughter. Jessica was an honors student – 6 feet tall, athletic, beautiful, artistic, a non-smoking vegetarian. One day, she phoned to tell me she decided to donate her eggs. She said, "Don't worry Mom. They told me there's a small risk of bleeding and infection, but otherwise they haven't found any problems." So she went one cycle without problems; in fact, she did it twice more in the next few months.

About seven years later, Jessica was diagnosed with colon cancer, a disease no one in my family had had. Two years later, after chemotherapy, surgery, and radiation, she died. She was 31.

After her death I published a 2008 paper in *Fertility and Sterility*, the official journal of the American Society for Reproductive Medicine, entitled, "Fatal colon cancer in a young egg donor: A physician mother's call for follow-up and research on the long-term risks of ovarian stimulation." I wrote it because I was shocked to learn that *no one had ever studied the potential long-term risks of egg donation, especially the risks of the high-dose hormones given to healthy young women.*

Unlike infertile women who are considered *patients*, egg donors are treated as *vendors*. When they walk out of the IVF clinic, no one keeps track of them. My daughter's death was not reported. The long-term risks of egg donation are unknown.

In 2013, we know that in addition to the short-term risks of Ovarian Hyperstimulation Syndrome:

- taking hormones increases the risk of several types of cancer
- infertile women who also get hormone treatment have an increased risk of cancer
- Jessica is not the only documented case of colon cancer in a young egg donor

- there are reports suggesting an increased long-term risk of infertility after hormonal stimulation.

But we don't know *anything* about the possible long-term risks for young women who go through this arduous procedure.

AB 926 Claims to Create Protections for Research Subjects

This bill will not accomplish this goal, because young women *cannot give informed consent* because information about long-term health risks doesn't exist. It didn't exist for Jessica, it doesn't exist now, and this bill does not require it. Before expanding the market in human eggs, let's follow-up current egg donors and collect their health data, so that real informed consent will be possible. I'm all for research, but not at the risk of the lives and health of more young women.

Appendix E

Testimony of Raquel Cool, Co-Founder, We Are Egg Donors, in Opposition To AB 2531

My name is Raquel Cool. I donated my eggs in 2011 to help someone conceive. Since then I and two other egg donors founded an advocacy organization called "We Are Egg Donors." We now have nearly 1000 members.

We support *informed* choice and *evidence-based information* to help women decide for themselves whether or not they want to provide eggs. However, we oppose this bill because, despite three decades of experience harvesting human eggs, the industry has failed to conduct the research necessary to support their claims that the process involves minimal risk to young healthy egg donors.

The industry claims OHSS is a preventable condition, but in 3 years I have seen only one case in which a doctor cancelled a cycle because of concern about protecting the donor. We are told that 10–20 eggs per cycle is the goal, and doctors know that greater numbers impose greater risks. But we routinely see numbers in the 30s, 40s, 50s and in some cases even up to the 70s or 80s. I can tell you, based on the experience of our nearly 1000 members , that moderate to severe OHSS occurs far more often than the 1% statistic that is being claimed.

Many of us have been overstimulated so severely that we have had to have our abdomens tapped with a needle to drain pounds of excess fluid.

We experience a wide range of side effects about which we feel we were *inadequately informed*, such as hard to reverse weight gains, acne and hormonal imbalances. More concerning are the conditions our members report such as polycystic ovary syndrome, endometriosis, cancer, infertility, premature menopause, etc.

Until those who want women's eggs are willing to support research on the consequences of egg harvesting so that we can make truly informed choices, we oppose expanding this market and ask you to vote against this bill.

Thank you.

Appendix F

Email to SFSU Faculty from UCSF Center for Reproductive Health, September 21, 2017

From: XXXXXXX
Sent: Thursday, September 21, 2017 3:17 PM
To: XXXXXXXX
Subject: Can you send this flyer about a unique opportunity for your students to earn $10,500 by donating their eggs at UCSF through you student listserve?

Hi XXXXXXX,

My name is XXXXX, and I am part of the UCSF Egg Donor Program team at the UCSF Center for Reproductive Health. We are currently recruiting women interested in donating their eggs for our program. Women must be ages 21–32 to apply. This is a great opportunity for students to earn money to pay for tuition or other expenses. Donors earn $10,500 for their donation and can donate up to SIX times.

Can you send the attached flyer to your students through your listerve? I am also happy to meet with you or talk over the phone if you would like more information on our program and donation process.

Also, if you know of places on or near campus where I can post flyers, please send me details on the locations I can post. Thank you so much!

Warmest Regards,
XXXXXXXXXX
XXXXXXXXXX
UCSF Center for Reproductive Health
499 Illinois St., 6th Floor
San Francisco, CA 94158
Cell: XXXXXXXX

Appendix G

Open Letter to President's Bioethics Commission from Fifty-Eight Civil Society Groups, December 16, 2010

December 16, 2010
Dr. Amy Gutmann
Chair, Presidential Commission for the Study of Bioethical Issues 1425
New York Avenue, NW, Suite C-100
Washington, DC 20005

> Cc: *Dr. Steven Chu, Secretary, Department of Energy*
>
> *Kathleen Sebelius, Secretary, Department of Health and Human Services*
> *Dr. Francis Collins, Director, National Institutes of Health*
>
> *Janet Napolitano, Secretary, Department of Homeland Security Tom Vilsack, Secretary, Department of Agriculture*
>
> *Lisa Jackson, Administrator, Environmental Protection Agency*
>
> *Dr. Margaret Hamburg, Commissioner, Food & Drug Administration*
>
> *Dr. Thomas R. Frieden, Director, Centers for Disease Control and Prevention Robert Mueller, Director, Federal Bureau of Investigation*
>
> *Dr. John Holdren, Director, White House Office of Science and Technology Policy*
>
> *Nancy Sutley, Chair, Council on Environmental Quality*

Dear Dr. Gutmann,

Thank you for this opportunity to comment on the Commission's recommendations on synthetic biology. We applaud the transparency and openness of the Commission's deliberations. Unfortunately this process has not resulted in recommendations that recognize the serious threats synthetic biology pose to the environment, workers' health, public health, and social justice.

The undersigned 58 organizations from 22 countries do not support the Commission's recommendations on synthetic biology. They are an inadequate response to the risks posed by synthetic biology because they: 1) **ignore the precautionary principle,** 2) **lack adequate concern for the environmental risks of synthetic biology,** 3) **rely on the use of "suicide genes" and other technologies that provide no guarantee of environmental safety,** and 4) **rely on "self regulation," which means no real regulation or oversight of synthetic biology.**

A precautionary regulatory framework is necessary to prevent the worst potential harms. This requires a moratorium on the release and commercial use of synthetic organisms until a thorough study of all the environmental and socio-economic impacts of this emerging technology has taken place. This moratorium should remain in place until extensive public participation and democratic deliberation have occurred on the use and oversight of this technology. This deliberative process must actively involve voices from other countries – particularly those in the global South – since synthetic biology will have global impacts and implications.

The Precautionary Principle Should Guide Synthetic Biology Regulations

The Commission's recommendations fail to implement the precautionary principle, and instead referenced the so-called "prudent vigilance" concept. The precautionary principle is recognized by international treaties including the United Nations Convention on Biological Diversity, the Cartagena Biosafety Protocol, the new Nagoya/Kuala Lumpur SubProtocol on Liability and Redress for Damages Due to the Transboundary Movement of Transgenics, and the UN Framework Convention on Climate Change. Although "prudent vigilance" is used as a guiding principle by the Commission in its recommendations, it is a completely new concept, apparently invented by the Commission without legal or policy precedent. When dealing with novel synthetic organisms that pose serious risks to the environment and public health, we cannot rely on a new concept with no agreed upon definition, framework, or precedent.

The precautionary principle often is mischaracterized as anti-science, anti-technology, or anti-progress. This is far from the truth. The precautionary principle, as outlined by the Wingspread Consensus Statement on the Precautionary Principle, states: *"When an activity raises threats of harm to human health or the environment, precautionary measures should be taken even if some cause and effect relationships are not fully established scientifically. In this context the proponent of an activity, rather than the public, should bear the burden of proof. The process of applying the Precautionary Principle must be open, informed and democratic and must include potentially affected parties. It must also involve an examination of the full range of alternatives, including no action."*[1]

Precaution does not derail progress; rather, it affords us the time we need to ensure we progress in socially, economically, and environmentally just ways. In the face of uncertainty and the potential for serious harm, synthetic biology will often require risk analysis. We do not yet know what the full environmental or socio-economic risks of synthetic biology are, nor has our regulatory system evolved to keep up with the science. That is why we need a precautionary approach.

Precedent exists within the executive branch to support the use of precaution. The President's Cancer Panel released a report in April 2010 on reducing environmental cancer risks, recommending that:

> "A precautionary, prevention-oriented approach should replace current reactionary approaches to environmental contaminants in which human harm must be proven before action is taken to reduce or eliminate exposure. Though not applicable in every instance, this approach should be the cornerstone of a new national cancer prevention strategy that emphasizes primary prevention, redirects accordingly both research and policy agendas, and sets tangible goals for reducing or eliminating toxic environmental exposures implicated in cancer causation. . ."[2]

This should be a guiding precept for the Presidential Commission for the Study of Bioethical Issues.

In October 2010 at the United Nations Convention on Biological Diversity (CBD), 193 nations unanimously agreed to apply the precautionary principle to the introduction and use of synthetic organisms. The CBD also recognized this technology to be a potential environmental threat in need of further review – particularly as it is applied to biofuels production.[3] This was the first time the United Nations addressed the issue of synthetic biology; ignoring this important decision would be negligent.

Lack of Environmental Risk Assessment

The Commission's lack of attention to ecological harms posed by synthetic biology is irresponsible and dangerous. The only ecologist to speak to the Commission, Dr. Allison Snow, raised serious concerns about the environmental risks of synthetic biology – but none of these concerns are reflected in the recommendations.

In her testimony, Dr. Snow presented four cautionary precepts to keep in mind about the ecological risks of synthetic biology and novel genetically engineered organisms (GEO):

1. *"We need to be very careful whenever novel, self-replicating organisms are let loose in the environment (intentionally or by accident). Many*

will do no harm out in the environment, but important exceptions could occur, especially if the GEO can multiply and become more abundant.

2. *Novel GEOs that seem innocuous or weak might evolve to become more successful when they start reproducing. Even if they are highly domesticated, mutations or unexpected properties might allow them to multiply in some environments.*

3. *Once these organisms are released into the environment, novel GEOs cannot be taken back.*

4. *Predicting which new organisms might cause irreversible harm can be extremely challenging . . . we have little or no experience with cultivating microalgae and bacteria outdoors, let alone new life forms that are entirely synthetic."*[4]

These points are mostly ignored in the guidelines.

The potential environmental impacts of the commercial use of organisms with synthetic DNA must also be examined. Many commercial applications of synthetic biology will undoubtedly lead to the environmental release of synthetic organisms – since it is impossible to prevent organisms from escaping from unsecured operations conducting activities described by some synthetic biology proponents as "akin to brewing beer."[5] More study also is needed on the risks of introducing synthetic organisms into the human body for biomedical and health-related applications, as well as on the risks posed by uses of synthetic organisms in agriculture. Since this technology is already being used to replicate pathogens, serious study of biosecurity risks is also necessary.

Even more troubling is the impact that synthetic biology could have on ecosystems and communities in the global South. A new "bioeconomy," in which any type of biomass can be used as feedstock for tailored synthetic microbes, is being enabled by synthetic biology. Biomass to feed synthetic microbes will be grown mostly in the global South, disrupting fragile ecosystems and exacerbating environmental damage from industrial crop production. Further pressure will be placed on land and water, which already are in short supply for food production, to produce fuels and chemicals that will be consumed mainly by wealthier nations. The Commission ignores these socio-economic and environmental harms despite the fact that already countries such as Brazil have felt their effects.

Unfounded Reliance on "Suicide Genes"

Despite the fact that "suicide genes" were explicitly described as having uncertain efficacy in Dr. Snow's testimony, the Commission relies solely on these and other types of self-destruction modalities as the main form of mitigating potential environmental harm. In fact, one of the main studies cited by the Commission in support of using methods to create "suicide genes" is still in an early development stage and has not been field tested.

Scientists who have studied "terminator technologies" in seeds have concluded that the process is never completely effective. They found that frequently occurring mutations allow organisms to overcome the intended sterilization thereby allowing those organisms to remain viable. Specifically, "suicide genes" and other genetic use restriction technologies (GURTs) represent an evolutionary disadvantage; selective pressures will lead organisms to overcome intended biological constraints.[6] Biological containment of synthetic organisms – which reproduce quickly, escape confinement, and cannot be recalled – is impossible.

Importantly, the UN Convention on Biological Diversity has mandated an international moratorium on the use of "terminator technologies" such as "suicide genes," and other GURTS that has been in place for the past decade. Reliance on an unproven technology that has been deemed unacceptable by 193 nations as the main method to "contain" synthetic organisms is irresponsible.

Reliance on a technology that will not guarantee biosafety or biosecurity and that has been prohibited by the international community is not a solution. Synthetic biology requires the strictest levels of physical, biological, and geographic containment as well as independent environmental risk assessment for each proposed activity or product.

Self-Regulation Amounts to No Regulation and Undercuts the Rights of Workers and the Public. Self-regulation cannot be a substitute for real and accountable regulatory oversight. Some synthetic biologists already have made several unsuccessful attempts at self-regulation. The second annual synthetic biology conference in May 2006, SynBio 2.0, was portrayed by proponents as "Asilomar 2.0," in reference to the 1975 meeting that proposed voluntary guidelines on recombinant DNA. At the 2006 meeting, synthetic biologists attempted to write a set of self-regulations intended to protect the environment and promote the field. This conference failed to produce serious results. Synthetic biologists were too concerned about promoting research and development to agree on even weak attempts at self-regulation.

The lack of open dialogue with concerned parties also contributed to the failure of the industry's attempt at self-governance. Civil society and the public, blocked from participating in these discussions of self-governance, issued an open letter to the conference participants. Signed by 38 organizations working in 60 countries, this letter called on synthetic biologists to abandon their proposals for self-governance and to engage in an inclusive process of global debate on the implications of their work.[7]

The current state of "self-governance" permits students to create synthetic organisms on campuses; and stretches of synthetic DNA may be purchased online, allowing laypeople to create organisms in their garages where, with no oversight, life forms not previously found in nature may be dumped down drains and flow, freely, into the environment.

The J. Craig Venter Institute and the Massachusetts Institute of Technology also attempted to draft self-regulations the following year in their report, *Synthetic Genomics: Options for Governance*. This report was limited in scope to biosecurity and biosafety in laboratory settings, focused solely on the U.S., and, importantly, completely avoided the topic of environmental safety. These experiences reinforce the need for real oversight to ensure that the real threats synthetic biology poses are never actualized.

The support of the Presidential Commission for the Study of Bioethical Issues for self-regulation undercuts the fledgling efforts of the Occupational Safety and Health Administration (OSHA) to put new safety requirements in place to protect workers using biologically engineered materials, nanomaterials, and novel organisms. The Commission's support for self-regulation undercuts the ability of workers to speak out and protect themselves. Becky McClain, a former Pfizer scientist, recently won the first lawsuit regarding a worker's right to discuss publicly the health and safety issues of the genetic engineering laboratory.[8] The Commission's failure to support lab scientists' basic right to know which synthetic organisms they may have been exposed to means those workers could become ill without being able to inform their doctors of the potential causes of their illness. There is nothing "ethical" about this kind of self-regulation.

Conclusion

The Commission's recommendations fall short of what is necessary to protect the environment, workers' health, public health, and the public's right to know.

We repeat our call for a moratorium on the release and commercial use of synthetic organisms until we have a better understanding of the implications and hazards of this field and until we have properly updated and effectively implemented public regulation of synthetic biology.

The time for precaution and the regulation of synthetic biology is now.

Sincerely,

African Biodiversity Network (Kenya)

African Centre for Biosafety (South Africa)

Alliance for Humane Biotechnology

Amberwaves

Asociación para la Promoción y el Desarrollo de la Comunidad CEIBA / Friends of the Earth Guatemala

Associação para do Desenvolvimento da Agroecologia (Brazil)

Biofuelswatch

Center for Environmental Health Center for Food Safety

Center for Genetics and Society Centro Ecológico (Brazil)

COECOCEIBA-Friends of the Earth Costa Rica (Costa Rica) Columban Center for Advocacy and Outreach

Columban (Missionaries) Justice, Peace, and Integrity of Creation Office (Australia) Development Fund (Norway)

Ecumenical Ecojustice Network Edmonds Institute

Environmental Rights Action/Friends of the Earth Nigeria ETC Group (Canada)

Food & Water Watch

Friends of the Earth Australia

Friends of the Earth England Wales and Northern Ireland Friends of the Earth Canada

Friends of the Earth Cyprus Friends of the Earth Spain Friends of the Earth Uganda Friends of the Earth U.S.

GE Free New Zealand Gene Ethics, Australia GeneWatch UK

GLOBAL 2000/Friends of the Earth Austria Groundwork/ Friends of the Earth South Africa

Human Genetics Alert (UK)

Institute for Agriculture and Trade Policy Institute for Social Ecology

Institute for Sustainable Development (Ethiopia) International Center for Technology Assessment Loka Institute

Lok Sanjh Foundation (Pakistan) MADGE Australia Inc.

Maudesco/ Friends of the Earth Mauritius Movimiento Madre Tierra (Honduras)

National Association of Professional Environmentalists (Friends of the Earth Uganda) National Toxics Network (Australia)

Natural Capital Institute Natural Justice (South Africa)

Oregon Physicians for Social Responsibility Our Bodies, Ourselves

PENGON (Friends of the Earth Palestine) Pureharvest (Australia)

RAFI-USA

Research Foundation for Science, Technology and Ecology and Vandana

Shiva (India) Safe Alternatives for our Forest Environment (SAFE)

Say No To GMOs!

Sempreviva Organização Feminista (Brazil)

South Australia Genetic Food Information Network (SAGFIN)
TestBiotech (Germany)

Washington Biotechnology Action Council

Notes

1 "The Wingspread Consensus Statement on the Precautionary Principle." Science & Environmental Health Network, 26 Jan. 1998: www.sehn.org/wing.html
2 Reducing Environmental Cancer Risk: What We Can Do Now. President's Cancer Panel, Apr. 2010: http://deainfo.nci.nih.gov/advisory/pcp/annualReports/pcp08-09rpt/PCP_Report_08-09_508.pdf
3 "COP 10 Outcomes." United Nations Convention on Biological Diversity. 2 Nov. 2010: www.cbd.int/nagoya/outcomes/
4 Snow, Allison A. "Transcript: Benefits and Risks of Synthetic Biology." *The Presidential Commission for the Study of Bioethical Issues.* 8 July 2010. Web: www.bioethics.gov/transcripts/synthetic-biology/070810/benefits-and-risks-of-synthetic-biology.html
5 Keasling, Jay. Amyris Biotechnologies. Testimony to the House Committee on Energy and Commerce hearing on Developments in Synthetic Genomics and Implications for Health and Security. May 27, 2010: http://energycommerce.house.gov/documents/20100527/Keasling.Testimony.05.27.2010.pdf
6 Steinbrecher, Ricarda A. *V-GURTs (Terminator) as a Biological Containment Tool?* Rep. EcoNexus, June 2005: www.econexus.info/sites/econexus/files/ENx_V-GURTs_brief_2005.pdf
7 ETC Group. *Global Coalition Sounds the Alarm on Synthetic Biology, Demands Oversight and Societal Debate.* 19 May 2006: www.etcgroup.org/upload/publication/8/01/nr_synthetic_bio_19th_may_2006.pdf
8 Pollack, Andrew and Duff Wilson, "Pfizer Whistle Blower Awarded $1.4 million," New York Times, 2 April 2010. www.nytimes.com/2010/04/03/business/03pfizer.html

Appendix H

Civil Society Letter to East Bay City Councils, 2011

September 27, 2011

City Council
City of Alameda
2263 Santa Clara Avenue
Alameda, CA 94501

City Council
City of Albany
1000 San Pablo Avenue
Albany, CA 94706

City Council
City of Berkeley
2180 Milvia Street
Berkeley, CA 94704

City Council
City of Emeryville
1333 Park Avenue
Emeryville, California 94608

City Council
City of Oakland
1 Frank H. Ogawa Plaza
Oakland, CA 94612

City Council
City of Richmond
440 Civic Center Plaza
Richmond, CA 94804

Dear City Council Members:

We are writing to raise concerns about the proposed second campus of the Lawrence Berkeley National Laboratory (LBNL) and the UC Berkeley Synthetic Biology Institute (SBI) that is being considered for one of your respective cities. Much of the research that will be conducted in this laboratory will be on the emerging technology called synthetic biology. Synthetic biology is an extreme form of genetic engineering that is attempting to create novel, potentially self-replicating artificial life forms from synthesized DNA. The risks this research poses to worker safety, public health and the environment are currently being ignored.

While some find promise in synthetic biology for manufacturing new products and helping us to better understand biological processes, it is an inherently risky technology. Synthetic biology research could result in enhanced virulence in existing hosts, heightened ability to infect a wider range of hosts, and resistance to antimicrobials, antivirals, vaccines and other treatment or containment modalities.

Laboratory accidents are much more common in the U.S. than most people realize and often go unreported. If there were an accidental release of engineered organisms in this lab, the health of workers, the environment and entire communities could be put at risk. Already, the current lack of adequate safety protocols and biocontainment within rDNA labs has caused serious illness and death. Since synthetic biology's objective lies in engineering novel life forms and products with the potential to interact with human biology and other cellular processes, we believe this research poses dangers (both from accidental and deliberate uses) unforeseen in the regulatory framework of standard rDNA research.

Therefore, before any decisions are made on a specific site for this new lab, we believe a comprehensive, independent and transparent safety and risk analysis capable of assessing these threats must be completed. It is simply unacceptable to allow the laboratory to self-regulate. Moreover, it must be ascertained whether such research is even appropriate near urban centers. Safety regulations and procedures must be created and tailored to address the novel aspects of this new science, including whistleblower protections and forums for workers to raise concerns, and the costs to any municipality of an appropriate public safety infrastructure must be identified.

Finally, independent regulatory oversight must be assured, particularly because both public and private entities will be operating at the lab. Every stage of this process must be open to and involve the public, including town hall meetings to discuss and address health and safety issues.

The Lawrence Berkeley National Laboratory and the UC Berkeley Synthetic Biology Institute must undertake the burden of proof as to whether their laboratory will be safe before any community can make an informed decision about inviting it to break ground in their backyard.

Sincerely,
Alliance for Humane Biotechnology
BioFuel Watch
California Coalition for Worker's Memorial Day
Center for Food Safety
Center for Genetics and Society
Council for Responsible Genetics
Friends of the Earth
Global Justice Ecology Project
International Center for Technology Assessment
National Injured Worker's Network
National Workrights Institute
Pesticide Action Network of North America

*If you have any questions or need any additional information, please do not hesitate to contact:

Jeremy E. Gruber, J.D.
President
Council for Responsible Genetics
jeeg@concentric.net
or
M. L. Tina Stevens, Ph.D.
Executive Director
Alliance for Humane Biotechnology
609-610-1602

Appendix I
List of Civil Society Organizations Concerning Emerging Biotechnologies

African Center For Biodiversity
www.acbio.org.za/en

Biofuelwatch
www.biofuelwatch.org.uk

Bioscience Resource Project
https://bioscienceresource.org

Center for Genetics and Society
www.geneticsandsociety.org

Center for Food Safety
www.centerforfoodsafety.org

Council for Responsible Genetics
www.councilforresponsiblegenetics.org

Econexus
www.econexus.info/who-we-are

ETC Group
www.etcgroup.org

Friends of the Earth
https://foe.org

Gene Watch UK
www.genewatch.org

GMWatch
www.gmwatch.org/en/

GRAIN
www.grain.org

Heinrich Boell Foundation
www.boell.de/en

Human Genetics Alert
www.hgalert.org

International Center for Technology Assessment
www.icta.org

Navdanya
www.navdanya.org/site/

Our Bodies, Ourselves
www.ourbodiesourselves.org

Prickly Research
www.pricklyresearch.com

ProChoice Alliance for Responsible Research
www.prochoicealliance.org

SynBioWatch
www.synbiowatch.org

Third World Network
www.twn.my

Glossary

List of Abbreviations and Terms

BLASTOMERES: The cells formed by cleavage or subdivision of a fertilized egg that constitute the embryo before cell rearrangement into distinct tissue layers.

CALIFORNIA INSTITUTE OF REGENERATIVE MEDICINE (CIRM): The agency created in 2004 when Californians passed the California Stem Cell Research and Cures Initiative (Proposition 71) into law. CIRM's, "Independent Citizens Oversight Committee" (ICOC), manages allocation of the $3 billion approved by voters.

CHIMERA: An animal formed by mixing blastomeres of different species. Interspecies chimeras do not occur in nature, but prospective twins in one fallopian tube can chimerize.

CIRM: See: California Institute of Regenerative Medicine

CLONING: The production of an organism, e.g., an animal, with the same set of nuclear genes as another organism. This can be accomplished by somatic cell nuclear transfer (SCNT).

CRISPR/CAS9: An acronym for clustered regularly interspaced short palindromic repeats and CRISPR-associated protein 9; components of a bacterial immune system that can be used experimentally to add, remove, or alter DNA sequences in any organism.

CYTOPLASM: The material contained within a living cell other than the cell nucleus. It comprises both the cytosol, the liquid-like substance enclosed within the cell membrane, and the organelles.

DNA: Deoxyribonucleic acid. The molecular basis for the perpetuation across generations of the primary structures (sequences) of specific RNA and protein molecules. The sequences are encoded in the DNA molecule by the order of its four different subunit types (A,T,G and C). DNA is not the sole medium through which biological features are inherited, but it is the most extensively studied.

DPTFR: Doctors, Patients, and Taxpayers for Fiscal Responsibility: The group organizing the NO campaign against the California Stem Cell Research and Cures Initiative

EMBRYONIC GERM STEM CELL (EGS cell): Pluripotent stem cells derived from primordial germ cells (PG cells). PG cells are cells of the developing embryo that give rise to adult gametes, eggs in females or sperm in males. EGS cells arise from PG cells placed in culture, and unlike the latter can divide indefinitely if not caused to differentiate.

EMBRYONIC STEM CELL (ES cell): Pluripotent stem cells derived from the inner cell mass (ICM) of a mammalian (such as a human) embryo. ES cells arise from ICM cells placed in culture, and unlike the latter can divide indefinitely if not caused to differentiate.

EPIGENETICS: Literally, "beyond the gene." Those effects, including reversible chemical modification of DNA and persistent environmental influences on determination of the phenotype, that lead to inheritable changes without altering the DNA sequence.

EUGENICS: The doctrine of improving the human species or sub-populations by utilizing biological knowledge, particularly genetics, to encourage the promulgation, by voluntary or coercive means, of favored characteristics.

GENOME: The set of genes characteristic of a given species. The human genome contains 20–25,000 nuclear genes; the genome of the bacterium *E. coli* contains around 5,000. The genome is not equivalent to the DNA content: only about 2% of human DNA consists of genes, though the non-gene DNA is also thought to have biological roles. In diploid organisms like humans and other animals each gene is present in two copies, one contributed by each parent. There is also a mitochondrial genome, containing 37 genes in the human, which is transmitted to offspring separately (though simultaneously) with the nuclear genome, via the egg cytoplasm contributed by the female parent.

GENE: A segment of DNA that specifies the sequence of a specific RNA or protein molecule in an organism's cells. The direct products of genes, including those that ultimately specifiy proteins, are RNA molecules. In mammals, including humans, most RNAs that serve as intermediates in protein synthesis are "alternatively spliced," that is, different segments are retained or discarded in an optional, regulated fashion. This means that the organism can produce many more different kinds of proteins than the number of genes contained in its genome. Every gene has many variant forms, most being consistent with a normal or healthy outcome, but some associated with dysfunction or pathology. Because of the context dependence of protein and RNA function in an organism's cells, gene variants that function well or badly in some individuals do not always behave similarly in other individuals.

GERMLINE MODIFICATION: Genetic modification of an embryo early enough so that germ cells, which will give rise to the new individual's eggs or sperm will be modified along with the somatic or body cells. The genetic modification will thus be carried forward to future generations, and not only affect the initially altered individual. This procedure is also called heritable genetic engineering.

GERM CELLS: The cells – sperm and egg in animals – that participate in fertilization to give rise to an embryo. The germ cells are also called gametes. They arise anew (with different combinations of gene variants) during development from fetal cells referred to as the "germ line."

HYBRID: A hybrid (between species or strains) results from the fertilization of the egg of one by the sperm of the other. The mule is a hybrid between a donkey and a horse, for example. Hybridization can occur in nature and can sometimes lead to new species, particularly in plants and birds.

INNER CELL MASS (ICM): The cluster of cells in a mammalian embryo that will give rise to the new individual's body, as opposed to the placenta and surrounding membranes. There are 100–150 ICM cells in a human embryo, the maximal number occurring at about 6 days of development.

***IN VITRO* FERTILIZATION (IVF):** A process of fertilization where an egg is combined with sperm outside the body. The resulting zygote is then implanted into a woman who has received hormone treatments to prepare her uterus to gestate an embryo. The children who result from IVF were colloquially known as "test tube babies" when the method was first implemented.

MATERNAL SPINDLE TRANSFER (MST): A procedure by which the the full set of chromosomes (arranged in a structure called a "spindle") are removed from the oocyte (prefertilized egg) of one woman and transferred into the oocyte of another woman. The constructed oocyte is then fertilized to form a zygote, the earliest stage of development. MST is a method by which the nuclear genes of a couple, in which the woman's mitochondria are impaired, is used to construct an embryo in which the mitochondria (and other nonnuclear egg components) originate in a second woman.

MEIOGENICS: The opposite of eugenics. The deliberate generation, or use of the product of inadvertant generation, of an organism with a deficency of the normal characteristics of its species.

MITOCHONDRIA: The cell organelles responsible for extracting energy from fuels (e.g., sugars, carbohydrates, fats) to drive the organism's metabolism or heat production. More than 600 proteins are necessary for the functioning of a mitochondrion, most of which are specified by the cell's nuclear DNA. The mitochondrion also has its own DNA genome, containing gene that specify (in the case of humans) 37 of its proteins.

MONOZYGOTIC: Two or more individuals that arise from the same fertilized egg (zygote), and therefore have the same nuclear and mitochondrial DNA. Identical twins, triplets, and quadruplets are all monozygotic.

ORGANELLE: A complex structure inside a cell, containing multiple proteins, and generally DNA and/or RNA. Organelles perform a limited set of specialized functions, in analogy to the organs of the whole body. Examples of organelles are the nucleus, the mitochondrion, ribosomes, cilia (for motility), and the Golgi apparatus (for packaging of secreted materials).

OOCYTE: An immature stage of a mammal's (e.g., human's) egg. Immediately after fertilization the egg attains its mature stage by dividing into two cells, the small polar body, and the large ovum. The pronucleus of the ovum joins with the sperm pronucleus. The entire assemblage is the zygote, which gives rise to the embryo.

OVARIAN HYPERSTIMULATION SYNDROME (OHSS): A medical condition with symptoms ranging from mild to severe, and occasionally fatal, that can occur in some women who undergo hormonal treatment to stimulate egg development and release. While there are rare cases that occur with no external stimulation, most instances are associated with preparation for IVF. Provision of eggs for research can also lead to OHSS.

PGD: See preimplantation genetic diagnosis

PHENOTYPE: An organism's observable characteristics or traits, including its form or morphology, its embryonic development, its biochemical and physiological properties, and its behavior.

PLURIPOTENT CELLS: Stem cells that can differentiate into all of an organism's somatic cell types.

PREIMPLANTATION GENETIC DIAGNOSIS (PGD): Analyzing one or more the genes of an embryo prior to implantation, or an oocyte prior to fertilization. PGD is typically done if the gene or genes in question are thought to predispose to disease.

PRONUCLEAR TRANSFER (PNT): A procedure by which the female and male pronuclei of a fertilized egg are removed and replaced with the pronuclei of a different fertilized egg. PNT is a method by which the nuclear genes of a couple, in which the woman's mitochondria are impaired, are used to construct an embryo in which the mitochondria (and other nonnuclear egg components) originate in the egg of a second woman.

SCNT: (See: somatic cell nuclear transfer)

SOMATIC CELLS: The cells constituting the tissues and organs of animal bodies, in distinction from the germ cells. Somatic cells are also often distinguished from the prediffferentiated and stem cells of the developing embryo and fetus, and the stem cells of the adult body. There are more than 200 different types of specialized (differentiated) somatic cells in the human body.

SOMATIC CELL NUCLEAR TRANSFER (SCNT): The transfer of the nucleus of a somatic cell isolated from a developed adult or juvenile tissue into an egg cell that has had its own nucleus removed. The reconstituted egg has a full (diploid) complement of DNA, so can be stimulated to develop into an embryo or potentially, a fully developed individual, without undergoing fertilization. The new individual will have the same set of nuclear gene variants (though not the same mitochondrial gene variants) as the donor of the somatic cell nucleus.

STEM CELL: A predifferentiated cell residing in, or derived from, a multicellular organism, that is capable of giving rise to indefinitely more cells of the same type and differentiating into certain other cell types. Depending on whether the stem cell can differentiate into all, many, some, or only one cell type of the adult body, it is, respectively toti-, pluri-, multi-, or unipotent. In addition to stem cells that occur naturally in embryos and mature body (somatic) tissues, stem cells can be induced in vitro from blastomeres (ES cells), embryonic germ cells (EGS cells), and differentiated somatic cells (iPS cells).

SYNTHETIC BIOLOGY: Referred to as "extreme genetic engineering," the endowment of an organism with properties not found in naturally evolved life-forms. This is typically accomplished by introduction of more than one gene into an existing organism, providing the basis of a biochemical pathway novel to its species.

T CELLS: Cells of the cellular immune system of vertebrate animals involved in recognition of foreign organisms and proteins, and distinguishing self from non-self tissues and molecules. T cells are distinguished from B cells, which as producers of antibodies are the major components of the humoral immune system.

TRANSGENESIS: The introduction of DNA sequences from one type of organism into a different one. The introduced sequence is referred to as a "transgene."

ZYGOTE: The cellular result of the fertilization of an egg by a sperm. The earliest stage of embryogenesis.

Index

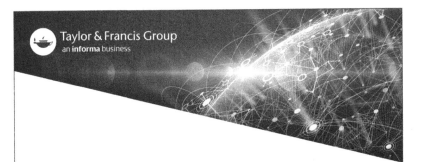

Printed in the United States
by Baker & Taylor Publisher Services